The Cosmic Origin of Life

Table of contents

Cosmic Beginning of Life

08-Introduction

10-Breaking Myths

14-Essene Gospel of peace

16-Crown center of creation

28-Origin of life in the salt water

35-Testing study on the Atomic interact.

41-The creation

51-Evidence of evolution

64-Creation in the salt oceans

68-Doctor Linus Pauling

73 Effects of ionization in saltwater

78 Rene Quinton - Dr. Linus Pauling

87 Path of divine consciousness

89 The heroic age of spectroscopy

105-Six- Creationist Conclusions

113-Part Breaking Myths

117-ARDI, the oldest human species

127-Beginning of life

131-Nostradamus (Accomplished) 2012
  new 2017 notes

136-Mystery of acquired knowledge

146-Maturity of the soul

151-Lighting of the self

156-Inner heritage

164-Alchemy (physical and mental)

169-The initiates

176-The concept of eternal life

180-The Darkness

184-Reincarnation

196-Short biography

The cosmic origin of Life

"Everything is determined, at the beginning, as well as the end, by forces over which we have no control. It is determined for the insect as well as star. "Human beings, vegetables or cosmic dust, all dances to a mysterious tune, they sang in the distance by an invisible Piper".

Good news from NASA's Hubble Space Telescope is that Einstein was right - maybe.

A strange form of energy called "dark energy" is looking a little more like the repulsive force that Einstein theorized in an attempt to balance the universe against its own gravity. Even if Einstein turns out to be wrong, the universe's dark energy probably won't destroy the universe any sooner than about 30 billion years from now, say Hubble researchers. "Right now, we're about twice as confident than before that Einstein's cosmological constant is real, or at least dark energy does not appear to be changing fast enough (if at all) to cause an end to the universe anytime soon," says Adam Riess of the Space Telescope Science Institute, Baltimore.

-Albert Einstein

"Stephen Hawking" -Einstein was wrong when he said: "God does not play dice". Examination of the black holes suggests not only that God plays dice, but that we sometimes confused by throwing them in place that cannot be seen. "-Stephen Hawking"

Howkins statement is certainly; formation of stellar bodies develops off, gases that are cooled on the other side of a black hole, and were not detected by the instruments in experiments performed by Hawking until April 8, 2015 new discovery

"Einstein said that if quantum mechanics were correct, then the world would be crazy. "Einstein was right- the world is crazy".

-Daniel M. Berger Green

Albert Einstein says that God doesn't play dice with the universe; Stephen Hawking says that God plays dice with the universe. Stephen Hawking says Albert Einstein is wrong.
Who play dice with the universe is the human being, to use the Boson, the primary energy of creation to manufacture nuclear devices for the destruction of their enemies and their own destruction.

Last minute received from Shannon Hall, Writer / April 8, 2015 9:30 a.m. ETA Yusef-Zadeh

The stars may be forming in the shade or exit on the opposite side of the Monster of the Milky Way Black Hole. Stars are born_in clouds of dust and gas. The turbulence within these clouds gives rise to worlds that begin to collapse under its own weight. The worlds grow hotter and denser, quickly becoming Proto stars, so named because they have not yet begun the fusion of hydrogen into helium. But Proto stars can rarely see. It has yet to generate energy through nuclear fusion, and no glimmer that this occurs; it is often blocked by the disk of gas and dust.

Therefore, under the new discoveries of science and its exponents, it is possible to say that a new link in human knowledge can be taken as true.

Spectral analysis revealed the chemical analogy between the stars and rose to the rank of certain substantial concordance of the Earth, with the remotest stars of the milky way and even distant galaxies. The display of material searchable in the cosmos unit, is the sublime lesson, historically the first granted us by the spectroscope with Kirchhoff and Bunsen. Cosmic emanations coming to our land are identical to that part out of this cosmic matter. The universe is and active mind that creates everything.

The creation is based on emissions which are parked in our terrestrial body to interact at the origin with salt water and then leave Earth to fresh water.

"The creation started and evolved into salt water"

We imagine that in the far universe are the clues to our origins, when I discovery that the energies that travel to our planet, are the courses of the beginning of life as local phenomenon, determined by the cosmic active mind emanations to our environmental.

The conditions to creative life are possible until now in planet Earth, as we perceive, but astronomers and history shows that other species, travel thru the outer space to earth, originating some different spices which can contribute to our origin's in the water bodies. Other species are proving to be in our planet earth, so we cannot discard that other cosmic planets support a different kind created life.

Printed in United States of America prohibited the total or partial reproduction of this work, by any method of electronic, computer, under the protection of the law within the limits.

ISBN-13: 978615685298 ISBN-10: 0615685293

Author: Edwin G. Pagan

Contributor: Joshua Burgos Pagan

8- Introduction

The creation sometime of myths in the history of humanity takes us away from the truth and leads us to creations and inventions to control well-hidden historical information that has revealed us today.

The creation is based on emissions which are parked in our terrestrial body to interact at the origin with salt water and then leave Earth to fresh water, our earth act as a mother pouch to create everything...

Quote: Our Liberty

"The world is a stage where everyone chooses the role that has to represent" Mahatma Gandhi

The advancement of the internal connection, the liberation of the axioms of that inner power, the inner dimension as a universal part. Usually the body of an enlightened is of a material rate that vibrates on a superior scale to the common matter. This attribute enables him to manifest qualities that the common being is not able to perform by will without the control of laws. His matures is a stage of development and spiritual initiations outside of material concepts. The liberation of material laws is another attribute, divine love, compassion, forgiveness, wisdom, sincerity, justice, peace and piety. The path of light that only they can receive and transmit to others, through the preparation of the being to understand the path of the mystic.

Jesus, for love of the father revealed the secrets of the creation and the laws of the active mind of the universe. Laws are manifest in the galaxies of the universe, and are the most sublime energies that the mind can grasp. The mental energy that will assimilate this link in the matter may enter a communion, in a latent state real and understand hidden processes of intellect. A true center of operations more perfect than modern computers which use similar sequences, the GPS of the human consciousness.

The millions of functions that handle these centers of communication since the first human being and all that late on this planet which we capture in our finite mind, other functions that are beyond our understanding and are beyond our control. Where in the universe arise plans that harmonize with a physical body that demonstrates a superior evolution to all known. We are a range of sums of phenomena; nothing in our evolution that matches something that compares or challenge this type of creation. A warehouse of construction materials is attracted and distributed where the specific cosmic subjects that will be part of our physical body.

If they knew that, within every being, there is a single world attached to a free pick by the divine nature that we generated as a human race. If this was understood, the freedom to create a new world within us would be the greatest legacy to humanity and a view of interior peace to understand the vastness of the cosmic universe inside us.

In my school memories about animals on the farm, never put much attention because all those days had that be with them where I grew up. To see the reality of the teachings was time wasted on me. Impressed me more a high pine tree that is Leiter pushed by the wind behind the classroom, where watched him out the window. In the Cup, it looked like energy fell from its dome and rose to the heights as strands of silver that dissolves in the air. That experience made me more knowledge that classes in the Hall at that age, a notion which I still have in my memories as vivid as the times I observe the phenomenon for long time. Out to the playground, I was engrossed lying in a small kiosk for snacking, watching the same phenomenon for a long time too. Being with the view raised by that high point, create in me a sense of giddiness that drove my body up to the navel and toes. A strange sensation of elevation came over me, sometimes ended up in a sort of partial dizziness.

"We have an interior dimension to discover"

Being must be like stem that bends to the wind, but not broken.

Lao Tse

10- Breaking Myths

When being enters in the energy control, it harmonizes with them, take control of physical environment in harmony with these cosmic fluctuations. Evaluation and progress of knowledge come into light of their true being and functions for overcoming these legacies of creation. Higher and refined power of God, in humans or things, power that being can master inheritance of God which the common people called miracles, power of divine emanations, where their laws are manifested. The enlightened beings that have access to that knowledge, those invisible powers travelling to this area of demonstration, where the only one studying them manifest is the human being. It is the main actor in the Theogony of God. It is the only one, the main actor in this scenario that may act per these manifest laws. The screening stages of improvement for those who do not have that legacy can emerge and seek a change in their lives. These attributes in this earthly plane are a replica of all the created laws, all sciences, all beings and their imagination. They are creator's bodies for the divine intelligence in greater or lesser degree; the universe is an active mind.

Gnostic Knowledge, knowing the reality internal divine laws and make them is the only channel. The divine energy where there is a name, only the essence of God. Dedicated to the fighters to make the light shine and being open his intellect to reality that dwells inside, "The Kingdom of heaven" carrying out the actors on this stage, that his mission is something transcendental. A quality follow teachings elevated and refined, for internal growth, peace, love, gratitude, to the God of our hearts. A magical power was left numb by the fall of man, is latent in the being, can be woken up by the grace of God, or the art of Kabbalah.

J. B. Van Helmont Leiden 1667

The human being who discover the potential of the Kingdom within his being, has the opportunity and duty to be a servant to the causes and divine laws of our father and creator, Jehovah, Elohim, Adonai and all the names that make the human mind aware, Yahoo, Yokahu and Ia-he-vau - I. The lost Word effluvium that generated the creation and the human being that is kept hidden in the most remote maze of being. Inside the field called Theurgy is that is breaks down the travel of the training of the human being, all the attributes in the creation of the interior consciousness and their divine original laws. Describes conscience as an attribute of the human powers to realize mature concepts, transform ideas, serve, being the reservoir of the miracle of life which affects God-consciousness, because he realizes. In the same way that you can debug your physical and spiritual being to rise to plans for more spiritual clarity, it can degrade you be lower emotions that characterize it. Want this say that it was created as well. That has the potential of rising is to it more spiritual, as down to it more gross of the manifestation. Because if God had to create pure things, so he did what is observed in people. It is a simple reason, guide our inner being to overcome the opportunities has debug their spirituality.

While most disclosure powers in the matter created by God and his laws, elements, their functions, the internal processes of divine creation, most will be taken to understanding the masses of their attributes in relation to the creation. Liberate the mind of fanatical ties, to advance in the mission that it is creating a universal mind of what brought us here to this earthly plane for which we were created. The universal laws of creation to which we have direct access, since being took knowledge of a divine concept of creation, have been manipulated for the benefit of institutions for their interpretation. History teaches us the various repressive attitudes that interpretation criteria have been imposed in a single direction to prevent the mental liberty to create... Purposes of positioning performances, acquire a notion of institutional for creations where in principle instead of uniting the races and Nations, serve to create a path of clash of emotions and different beliefs. The reasons that are glimpsed behind these historical accumulations have its creators and their purposes. The path of leaders to achieve the power and dominance over other cultures at universal level. It is not necessary to cite examples; modern society is a mirror of these assertions. The universal laws of creation do not obey to these routes of temporary powers. The notions of every human being are very diverse, depending on the influences to which it has been exposed, but freedom of thought leads us to the truth and liberty. The rest is simple to realize and seek truth internal within each being. The Kingdom, the same freedom that we owe to God, not to the slavery of the self by the self. We are the only ones that we transmute and create the magic of its manifestations, it is our legacy and the joy is to transform and freely create.

It is very suitable to reserve this space for copying a fragment of the "Essene Gospel of peace", left by Jesus and rescued in the caves of Qumran by the Essenes and zealots (zealots - the military arm of the Essenes) when invaded by the Roman Empire in the year 65, the Jewish territory.

Note: The way the scholars in religions try to present Jesus as an incarnated GOD, was the creation of Saul Paul. Paul was the nephew grandson of Herod the Great, of Arab Nabataeans extraction, neither Herod the Great nor Paul were Jews, claiming to be to aspire to the throne of Israel.

In the apocalypse sent from Egypt to John, Jesus denounces them and asks Simon Peter to protect the manuscripts of those who claim to be Jews and were not in clear reference to Herod the Great and his great-nephew Saul Paul.

The discrepancies between Peter and Paul arise in the Synod of Jerusalem. Grew up under Tiberius Alexandro (46-48 AD) due to a great famine that occurred in the country and thus refers to the many incidents that are mentioned later - this happened at the synod of Jerusalem where Simon Peter and James; were denounced by Saul Paul to Tiberius Alexandro who send both to be crucified. This legacy brake some myth created.

14- Essene Gospel of peace Jesus teachings by John Writings

And then many sick and cripples were Jesus, asking: "if all know you, tell us why you suffer these painful pests? Why are we not whole like other men?

Master, heal us, so that we become strong and not have to live any longer in our suffering. We know that your power is healing all kinds of disease.

Get rid of Satan and all his great evils. Master, have mercy on us".

And Jesus answered: "happy you who are hungry for the truth, then you will satisfy with the bread of wisdom. Happy you that you call, because you will open the door of her life. Happy you that you reject the power of Satan, because you give to the Kingdom of the angels of our mother, where the power of Satan cannot penetrate.

And they asked with be wilderment: "who is our mother and what his angels are? And where is his Kingdom?"

"Your mother is in you; and you in it. She gave birth to you and it gives your life. It was she who gave your body, and you'll it to her again someday. Happy you when you arrive to see it, as well as his Kingdom; If you receive the angels of your mother and earnestly fulfill its laws.

Verily, that whoever does this will never know the disease. The power of our mother is above all. Destroy Satan and his Kingdom and has Government on all your bodies and all living things.

"Blood flowing in us is born of the blood of our <u>earthly mother</u>. His blood falls from the clouds, springs from the bosom of the Earth, murmurs in the mountain streams, thoughtfully flows in the rivers of the Plains, sleeps in the lakes and becomes enraged powerful in stormy seas.

"The air we breathe is born of the breath of our earthly mother. His breathing is sky-blue in the heights of heaven, whistle in the summits of the mountains, whispers among the leaves of the forest, flies over the wheat fields, dozing in the deep valleys and burns in the wilderness.

"The hardness of our bones was born from the bones of our earthly mother, rocks and stones. Is stand nude to heaven at the top of the mountains, are like giants that lie dormant in the foothills of the mountains, as raised idols in the desert, and are hidden in the depths of the Earth.

"The delicacy of our flesh is born of the flesh of our mother

Earthly; meat matures red and yellow in the fruits of the trees, which feeds us in the furrows of the fields.

"Our intestines are born from the bowels of our earthly mother, and are hidden to our eyes as the invisible depths of the Earth.

"The light of our eyes and our ears hear are both colors and the sounds of our earthly mother, that surrounds us like waves of the sea fish, or as the air swirling the bird born.

"Indeed, you say that the man is son of the earthly mother, and of she received the son of the man all its body, of the same mode that the body newly born is born of the breast of his mother."

Verily I say unto you, that you are one with the earth mother; she is in you, you in it. It you were, in her live and her new returns. Keep therefore his mild, because no one can live long and be happy but he who honors his earthly mother and complies with its laws. Your breath is your breath, your blood, your blood, yours bones their bones; yours flesh their flesh; your bowels their bowels; your eyes and your ears are their eyes and ears.

"Indeed, you say that if debases of meet a single of all these laws, if damage one only of the members of all your body, you will lose forever in your painful disease and would be the cry and grinding of teeth." I say to you, unless you follow the laws of your mother, not you can in no way escape death. And the laws of his mother, who embraces him, embrace his mother also.

It will cure all your pests and it never get sick. She will give you long life and will protect you from all evil; fire, water, the bite of venomous snakes. Well since your mother gave you birth, preserves life in you. She gave you her body, and only she will heal. Happy is the one who loves his mother and lies quietly in her lap. Because your mother loves you, even when you give it back. And the more you'll love if you move back to it? Verily I say unto you, that great is your love, larger than the largest of the mountains and deeper than the depths of the seas. And those who love to your mother, she never them leaves. As well as hen protects her chicks, as the lioness her cubs, as a mother to her newborn, thus protects the earthly mother to the son of man from all danger and harm.

16-Crown center of creation

The process of emanations from the Crown to the realm, passing through the stages of creation and evolution, where the divine energy built in is its cockpit, to express his performance, in accordance with the subject and the divine creative force. As it is up being down, a perfect complement to the manifestation in this earthly plane, creating interact with energy that I create it. I should give more insight to things that have never been in contact with this medium, see the truths of previous civilizations, which, for reasons of historical debacles, the struggles of peoples, is the story that we have linked to the civilization. The mysteries involved in early civilizations, were breaded by myths and its true meaning. It was subverted by accommodating statements, which then represented what was beneficial for the institutions of Government and control of the masses, a power behind the government controls. Especially those promoted under Government controls and idealistic currents that have been inserted in education, to ensure their interests, in separate Nations. The best data are in specialized books. There is a reality, which moves into the surrounding walls to that story and has buried thousands of narratives that lead to disguised concepts. Do not offer a Verity of front and real presenting himself to mankind. As hidden it is that nobody realizes, and those who discover it are put in warning of the price that must be paid to cross that threshold of control and protection.

God loves the harmony, which use imagination to create is your enjoyment. The great teachers were creators of works that transcend the imagination of them others beings, that paid by those creations and is believe gods by own them. Only geniuses have bathed in God's grace and his soul was enhanced by their performances, apart from the material. Linkages with the evolution and creation that are usually discussed and produced theories to where I have could reflect, is in relation to the studies exposed on this issue. The physiological form of the evolution of beings, created kingdoms, our curiosity to know how everything works. Addressing this maze is an exciting adventure that we must undertake all. Not must expect that others built their mansions and castles within us. We are the masters and owners of our interior mansions.

The insulation of the creative force that emanates from the outset, remained protected inside a natural cockpit or transmembrane, closed to the influence of the foreign material forces in the salt water of the seas and oceans of the planet. The thalamus mother gland, as formed from the first transmembrane in the creation by the influence of energies emanating from the cosmic part as vibrations, that starts the life in our planet, as a permanent universal creative law.

The tests are documented by Claude Allègre experiments on the subject, in this book (A bit of science for everyone). Other studies that apply to the subject will be given to know, with basic laws of the origin of life.

Interact in these energies created a callous structure isolated, where he developed a neural center that interacts with the other endocrine glands generated in the brain. The awakening of this hub of activity gave rise to what would be the beginning of the creatures and the human being, in the salt water of the first Lakes or oceans surrounding the first continent of **Pangea.**

As above so below, a small part of universal mind hidden inside us.

Is in that Center where is encapsulated the paradise, and the intelligent force of what generated the life in the Earth. From some external source, traveling from upper or outer scale of vibrations and waves of energy was projected into the matter by the primary laws of emanation. The energy that the human mind seeks to unravel their origins and qualities, which keeps us busy in search our real origin. Ancient thinkers and analysts of the qualities of the being and the matter was a challenge to give with the attributes of the creation. The functions are studied per their internal manifestations and outcomes of observations made of the phenomena. The systems created by the metaphysicians and mystics had bases in an introspection of the faculties of internal perception. The association of emotional, the fluctuation of energies and the sensations that are lived in them, different states observed. Pure creation science than we are bodies inhabiting a planet, that we came here from somewhere like a new child and spent a period and disappear. Now comes, another rebirth and wonder where all these things arise. Are we the same as we started some time ago and returned? We realize that we already have a maturity, since we opened our eyes to the world. There must be all sorts of issues, relating to concepts studied. It is part of what I will try to cover my experiences or my way of viewing this topic. We come into the world with a family host, inherited legacies of genes from our parents. This chain of events recorded in the universal conscience of things, remains latent in the things by repetition. Evidence of repeated cycles of existence, they recreate qualities inherent in a previous life demonstration. That combination, our being, is defaulted to a development of repetitive qualities in the structure of development. We realize and recognize most of the things that surround us at all levels by association, we began a nearly instant relationship, and many times we recognize that things have lived them before. We sense a spiritual heritage that accompanies us, once we have training can continue develop skills and knowledge to personal, individual level. At first, we are not trained by our teachers to wakeup that part of our internal wisdom legacy, our parents were not trained to recognize that natural inside paradise.

'The bridal chamber' thalamus - Eva and Adam were prisoners of foreign influences. In principle, the paradise, then degradation of the Kingdom, this involved the creation of the self. Divine and sacred, baptism where mingle, holy water and the divine energy. The energy that takes the structure or logos of every being and everything, the essence of God interacting with the matter that creates. Penetrating that legacy, released to respond to the legacies of the intuitions of spirituality that flow from the cosmic consciousness of God, to the understanding of divine principles that emerge in our consciousness.

Our physical expression is real and we carry it everywhere, it is our physical being that moving shell returns to Earth as a new kid, to be born again must reincarnate. Who can deny this reality, denies the same God and himself. We are the stage where God manifests, the only one who realizes because he forged us to be their reality. Endocrine glands that accompany brain have a specific task to determine the frequencies and shares assigned to each component added to its skein of the nervous and physical system. Functions are adapted to the way that develops the individual and everything manifested in his Kingdom, and the sensorial experiences of their development. It is imperative that humans take a moment and ponder on the functions of your inner body, where functions are not stopped, everything happens as a synchronized mechanism. We leave to those impulses of attention, since it is something that belongs to us at first glance. At school, we taught the processes of voluntary and involuntary, that border is few, those who impress you. The lungs breathe with a rhythmic sequence and for periods of time. The heart beats at a pace, to supplement the body's basic needs as well as other organs. Many of that process can be interrupted by a human for short periods of time. We can look at where we fancy, walk, lift your arms, all that is mechanical in us so dominate at will. But not can't, let's think, grow, stop the imagination, leave of breathing, stop the heart, make that the body leave of create cells; the heat of the body, circulation, fail to hear. But without realizing all this happens like clockwork, nothing stops him as their functions are not interrupted by the cessation of all processes or atrophy of their components. Meditate on the creative energies that flow through our material bodies, require a control that partly impress us.

We can enter its depth to bring it to reality in which we live. We can accommodate the communication; become the real idea that is projected in our consciousness. More emotions causing us can be communicated to others. The depth of perceiving these, something mystical, that overwhelms be total. It is perceived as a channel where this notion is projected up to our being. While the material part follows its course, they stop outside the required sequences. You are incurred deficiencies affecting the normal functioning of the body, but nobody gives us rope. We realize in those details that we don't have the power to stop creation and evolution. We cooperate with it and leave us to influence its processes to harmonize them, achieve a greater vitality of its functions. We take food and these are added to the already accumulated energies, supplemented each and create additional per items that we add to our structure. The same applies to the information we process mentally, as it adheres to us, be thinking such as soot or a layer of what we accept as true. Have the certainty of that the system automatic with which were gifted by the nature, is potential and the only that can assume the control of that machine created in your all, is cosmic energy living in us. We must think with freedom ideals that are taught to us, not to succumb to idolatry.

Small account of being their internal functions

Awareness is not another thing to realize, discern an intuitive knowledge of what we are, from the observation of the combined cycles than is shown in that function. Another function of consciousness is that it can flow through the internal matter to realize the functions. Simply view our internal body and educate that function to recognize organs internal glands, their functions and harmony with each other, give us a map that is learned about the human body. Rather than learning from its internal components to view the harmony of the set of functions that take place without our control, would make us masters of our inner being. An education by repetition of events and maturation of behavior accepted, matured and taught to those who share the same family environment. The accumulation of data that gives us evidence to situations for which we have not passed, but imagine them and we have a certainty by imagination than they are. If we talk about a Red House we can imagine it, but not a definite form we associate with that concept, if we described wood imagines a model of something found or seen before. If we are given a description of qualities, we reach a mental picture of the House as we described. By Association accept education, from other sources. By some unknown inspiration, we imagine things and events, never known, we mature them in the imagination, and after a reasonable period, capture them into something real. That period created internally, we can manifest on the physical plane, an imaginary work that is to create. From my own experience, we keep a vivid reality that is recorded within us as a movie of the experiences of what is perceived externally and how they affect us. We realize that we can create inside and outside of our physical environment, it is a dual force complemented. Occasionally pass through sensory experiences that put us to think, imagine an unknown world that we cannot penetrate, only imagine what happens. An ocean of experience accumulates in our daily living that we'd like to unravel its provenance. Many times, access internal frequencies of any dimension, which is not common with the material life. Emotions and spiritual elevation, is as if an inner channel in our consciousness that lifts us opens to a dimension never experienced. These parts of energies that commune with our extra sensory experiences are catalogued in a universal or cosmic mind where everything emanates from the divine mind where everything is contained.

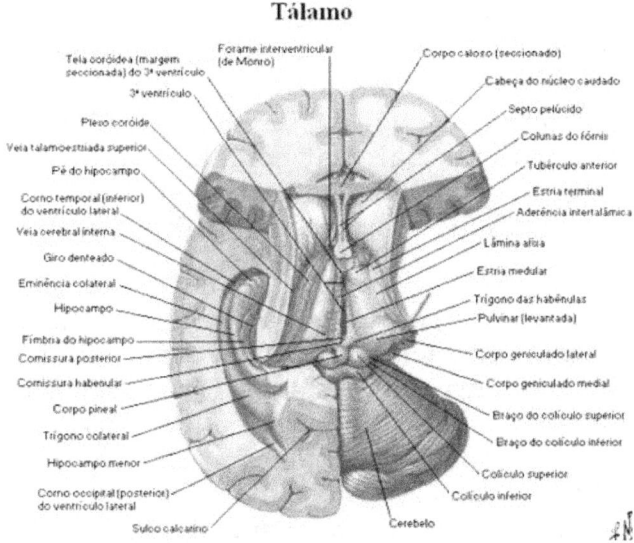

Harmonize ourselves with these frequencies gives us a certainty of consciousness functions and internal attributes that allow us to perceive that dimension out of this plane.

Our endocrine system and lower thalamus forces View our inner self Start communication over networks that emerge from its Center in communication terminals or nerves in the central nervous system. It's see is an extension of the systems of neurons which graduate a constant combination of materials to maintain the flow of energies of regeneration or construction of new sectors by impulses. The central nervous system is the channel of communication and signal where it is adjacent to the skin, which serves a range of reflections and exchanges of energy, interior materials that travel to fertilize or harmonized with internal processes. We are a large magnet of energies that constantly combine to complement material aggregates that added in its interior, through the consumption of food materials. Them added of energies of the aspiration, liquids, and compounds that enter to the system, of that aggregation of energies is adhere to which invade our inside of different sources external that combine what is the man. A balance supply of foods and oxygen and other untouched combination formed inside this gland as a mysterious law.

A network of veins, capillaries-arteries, valves and communication tunnels bypasses our internal system. All building materials that the body has acquired for construction and aggregation of materials are transported through it. The secret that many are not allowed as in the process of the body, the energy of creation separates each component for their individual qualities and directs them by this network of internal distribution. Precisely to the right place, in the amount needed, with an internal intelligence of what each thing and what will be their role. Before adding the function, which should be replaced, reach the exact spot, start its functions without further instructions. A channel of sensations for each differential function; each function has a sphere of feeling different and separate from the others, gather at a given point that feeling is attached and added to the other forming a circuit, demonstration and then the cycle of many others that are added to the primary feeling. But we do not take knowledge that is more complex than a computer, is in the Centre of the head, a small Center of operations, called the thalamus mentioned in the books of science and endocrinology. Its true performance is a myth for most those who do not enter higher or scientific studies. I've not familiar with previous, or (Darwin's theories), I've only heard in high school. What is something intuitive partly paid by unbiased readings, so far and knowledge acquired in the esoteric subjects. (I include a different information that demystify Darwin Theory in this book) Intuition is a channel that moves in subconscious memory, is a function that I termed the channel of communication of the material with the spiritual. The border where they meet the divine forces with the physical ones the being sensed and made the essence of God; it succeeds thanks to those domestic sources of communication where it is perceived. If we pay attention to all internal functions, we can understand god's creation.

The other part is a small maturation of how control, the flow of internal energy by concentration the attention in specific, areas setting the attention aware and viewing the internal functions. If we focus our attention to the muscles of our body, we realize how these automatically reacts, the less need to apply duties to a body reaction to carry out any movement you want to activate. The reaction is immediate and the reorganization of millions of muscle fibers complies in exact precision to the action that should be performed. Another way to train the internal system in which we can influence, is concentrating extra powers, do repetitions of functions decided outside the automatic reactions learned by body system. We can build our system of control of certain parts of the cycles depending on some areas, if we know the processes that we must apply. This was wisdom of many Mystics who managed to transcend functions to do what people would call miracles. Directing the attention by means of viewing processes, at the same time which applies a portion of breathing that adds energy, which is present in breath, so that the material body gets more concentrated component to interior construction process. It's like cooperation to the processes of maturation and aggregation of a natural process wrapped in a premeditated notion. Is a process learned of the wisdom ancient schools, to harmonize the energy of the body? Within this exercise an extra notion of something that only prepared Mystics filtered. A notion of a spiritual force greater than all that matter eventually processed.

The steps to be internalized the mind of those higher functions. The primal impulses arise from the need to create them the environment and its components. The first egg of creation, bringing together cells of proteins that are beginning to occur by the thalamus; (the first egg of creation) generated process inmates of attraction or magnet inside what we call the aura, converts it in impulses or law of attraction, other elements which contribute electrons and complement their development. That quality or notion of a matter and its components are both a divine intelligence that acts towards creating and isolated her in a cabin, where started the control and the stages of creation of the self. In a voluntary process is they can influence internal currents, which emerge from the neurons to travel by the system of the nervous complex, which clothes the full functioning of our interior.

In a principle that was a reality for the human, with their attributes is stated such which is. These are legacies that do not recognize today because the processes have become automatic copies of the same. Memory created from these functions is already aware and distinguishes one from the other. His appearance was in the first bodies of salt water, the first reactions to the released and attract the energy. The first cells created a kind of magnetic fence where detected to her around other powers that were not allied with theirs, to reject what was not harmonious with their structures. This addition was creating mixed reactions, because you interact with other creatures that were developed in a common sea water environment. Time created a kind of magnetic field that responds to what we call aura. This is a field that responds to powers that act within the functions that is carried out internally. It is endowed with qualities of detecting other energies, at some distance from it. Reactions are activated to protect of what is coming. So, necessary was in development, many species create rejection systems, the more sophisticated can be imagine. Electric shock, camouflage, extraordinary sounds, or directs attack, with attachments that the situation believes in their structures for defense. If you look at its structures were not born with these elements, these were created by the need to overcome conditions of their environment. This is the nature that we see in another species. The magic of transforming matter into control laws is a mystical and divine legacy of our Creator. Since we assume the functions or laws and his new creations, started interacting and the magic of transmutation of matter and the divine forces of being inside. The power that we have for heritage is a divine legacy, we are holders of the divine mantle of creation.

28- Origin of life in the salt water

The divine particle discovered. The CERN laboratory:

In statements by the year 2012 is described in the news that finally has been given with the particle that gender creation from an infinitely small particle or particles of what everything was generated. Archimedes possessed mastery, mathematical and intellectual strength to demonstrate 250 BC, the time and the concept of space. The Egyptians also took part in the accumulation of this knowledge. The infinite interior lighting is the same as the God particle. If we cannot conceive within us as its creation, is not another thing that demonstrates the energy that moves us. The way in which the mind or the subconscious are impressed by this energy, is equal to the reality of cosmic energy. If it true awareness of this dimension, material testing, the only thing that shows is that we are part of it and the consequence. If not are capable of give us has that can describe what is or discover their last particle, at least is more than a printing physical of what is. The depth of the soul, human can transcend to that dimension.

Dr. Linus Pauling along with Rene Quinton, who before he presented conclusions on the same subject in the year 1904. Linus Pauling: Postulated that the human being has 118 elements of the periodic table in your body, which are present in water body's salty seas and lakes. Also, announced that, cell life, originated in the Precambrian, 3,800 million years ago, in the salt water. Another finding is that human plasma is like sea water, into its components.

Note - with the variation that human beings now consume fresh water and should have a slight variation in composition, although it from consuming processed salt.

The geometry was the first expression of the divine mind in its first scent of where the spark of creation originated. -Kepler – 1619, the order of a figure and the harmony of a number, evoke all things - Giordano Bruno - 1591 Johannes Kepler GOD geometrized. The fifth essence per Plato was an Egyptian education where the world is in its most minute particles composed of triangles grouped into five elements or the quintessence. Kepler corrected this proposal of Plato and enunciated that God entered exactly five bodies between the distances of the orbits of the planets. With that discovery argued - "I've stolen Egyptians the five vessels of gold, to erect a sanctuary away from the borders of Egypt to my God. The five figures intertwined by the planetary forces that compose the octahedron. "Within them is the divine energy that only they handle."

Archimedes had a great mentality and demonstrated the same principle in geometry by equations. In any case, behind the cabal mental Alchemy, gives the idea of the fall of the upper bodies, was caused by the fall of Adam Divine degeneration to be created. Paulus Ricius 1516. The human being has not stopped his hunger to simplify or theorize God. The scholars, who meditated before us, obtained a deep understanding of the primary cause. Current beings are floating in the diatribe from whom or which best describes God. If the clear majority of the humanity accepts that what not can understand and communicate of them internal qualities, it tops in what the cosmos wraps their functions, that same cosmos is for them beings the God of their consciences. Scientists to reduce those functions to the microcosm, be common only sees the macrocosm of its manifestations, and those scientific deities and cosmic forces catalog atoms, electrons and protons or particles, in the way that is holed of God. Adopted concept that altered the order of things and take over the world raising to myths of incarnate gods.

At the end, will be that, his concept would be to realize that we are only ones we do, if humans did not exist, I said, God would not be a reality. God is the mind that does it. The primary energy manifested itself in the human being but is not the human being itself. The human being is just a channel of expression of the primary force, which covers the entire cosmic universe in a real manifestation of it, but the only inside sublime energy channel can react to its attributes.

Material creation

A shadow of what started, the volcanoes that constantly interact with created matter, a few lakes where the basic elements that clashed with the mass of matter that was the beginning of planet Earth were concentrated. They have already existed a material conformation where was expelled until it reacted to the laws of attraction and repulsion and is set into an orbit in the solar system. The reality of occupying an orbit in a complex of galaxies in the universe, gave us that quality of being what we are. Which, in this small conglomeration of materials and energy, have been concentrated in this atmosphere, is the demonstration that our senses are not matter on the step of evolution where we position ourselves? Streams of matter and all its components from the simplest to the most complicated merged into States that all we see daily.

The mysteries within these are the challenge of human intelligence. Starting from basic compounds there began to form with the eruptions of hot lava, creating the water that is condensed for long periods of time. Layers cooled and the holes created by the rock that was not sealed were provided as liquid strainers stuff, where they concentrate minerals diluted in the silt, resulting from these leaks and mixes with sediments. Passing of time, the first simple bodies, demonstrated in the broths, which are accumulated and recreated the first cells. With reactions and pangs they were dragged to the currents of water that accumulated in basins left by the formation of the rocks that cooled in the telluric activity of the Earth where large amounts of energy were released to outer space. On the seabed is to start off the first cells to form per the marine soup that engulfed them. This conformation is that the first characteristics, given to the reaction that arose individual entities, per the combination of matter and energy that were exposed come in a play of forces and energies. The internal components of matter are manifested in a reverberation of atoms, energy of primary organization. Layers of matter are accumulated to be already organized into something with its own characteristics, as bacteria that gave base to fungi and ferment with chlorophyll to absorb the sunlight. Energies that emanate from the area attracted by these reactions focused first on the waters, (H_2O) - (O_2) - as it was the source of attraction and combination, given that elements of the powers combine materials of different categories of material. The interactions of solar energy, the Moon, the emanations of the planets, the cosmic energy of the galaxies that broadcast its attributes was as the planet Earth.

The magnetic forces of the planets that keep it in orbit influence in the nucleus of cells created to establish identical to the planets rotation dynamics. The centrifuge forces of attraction towards those bodies. The continuous change of them, forms is must to these influences of forces foreign in their atoms of it which is manifest. Within a layer of electrons related matter, opens a rift created by the same reaction of these energies. Its structure regroups in layers, with different degrees of Congregation and attracted other sources energy that balances its structures, in the same way that it destabilizes, this principle also looking for a standby force. The process of creating layers to isolate processes makes them, Interior and exterior. The dynamics of this sequence creates an interior cabin, where the energy is blocked between two walls. The rebounds of energy released from that combination, strangled in these dense layers make function of triangular spaces, where matter begins to create extensions of the same. It is there a natural reaction that first created cell gathered his forces in the center of his being or control vibration; vibrated along with the light-energy concentrated in it. Is moved by impulses of attraction towards the source that it projected by its own nature, bounced by reflection (a reflection it same that a mirror) doubling a radiation of their same quality. Shock produces waves of other variables that come into play in the harmony of what is interacting; part of this energy as released.

Out of their orbits to be attracted by other atoms are in imbalance of power and projected outward. Which fail to be captured by other waves and nuclear orbits, for any behavior is drawn inward forming a triangle, combination was trapped in the undulations walls of vibration in a circle, where power was concentrated and irradiate and established a balance reflection. The first notion of circular energy that affected being. The creative energy takes shape in what would be a new created entity. Comes to know his source to react to the dimensions of the inner movement, their ability to react to movements and downloads; they trained its continuous and immediate reactions. With the concentration of primary materials, in this case sodium, potassium, oxygen, hydrogen, carbon, which were presenting in saltwater why by concentration? Create the proteins that give rise to the first membrane at the current position of the Earth, or as an inherited attribute of its previous orbit where snapped. He clusters of reactions gave it base to the first transmembrane that created a matter specific, the proteins, so his reaction of remain active is concentrate in she, the proteins you gave the support to survive. The primary cell formation, was androgynous, in power of two polarities which were separated by the same laws of reaction, each adopted qualities react with each other, the attraction of the one by the other and force manifested in undulations of light; is rejecting the one to the other by primary chaos; What gives rise to the polarity aligned per their chemical qualities that is activated by the flow of magnetism. It penetrated the first membrane which formed matter, in its motion, was repulsed by the repulsion of counter power; separate by space is the attraction between the nuclei created. The fusion of attraction forces sets the first reaction of what would become the first cell of the matter that gave base to the two lobes of the thalamus. In this case am oriented to describe what is the principle of the human being, not ignoring the creation of them others species and kingdoms that are produced to the same time, but with emanations of different vibrating rate, acting with matter and different conditions. The emanations internal recreated the first matrix of the membrane that is formed. The combination of these emanations is mixed and fecund it; of there was born the trine force of the love, interact, the harmony and by involution (a universe in miniature, the central force, vibrating with the laws of the primary creation). A Cosmic harmony was born, the first glow of a cosmic Symphony that encompasses the phosphorylation of the white light into seven basic colors. Arise the first harmonic scale, which gave base to the projection of the elements in the cosmic space. The verb, the first law that completed the first divine action in relation to the triangle, parenting - mother with the two human powers)-created the missing dimension. The divine manifestation of primary energy interacting with the vibrations of the infinite mind, (the energy generated from the source of primary vibration, which reached this distance created a uniqueness in the universe) which took consciousness of itself, in its first conformation. Emerged the material dimensions of the being, dimension which gave shape to a reality that at the same time reflecting its presence, arises the creative perception of the Logos.

Where did not exist before this notion of existence, was vomited to the material manifestation where it started interacting divinity and being created, high, low; light and darkness; right, left; the notions of divisions who graduated the powers to give the first active mind to the creative energy. To express her, it was degraded to the point to get to know and become aware of it, but as a dimension never manifested. A dimension in which trained itself to flow with this creative freedom, the Ionic permeability to both sides induces it to a change of transient, generating an electrical impulse. An arc of meeting in the form of mantle that engulfed commodity, in reactions of forces of attraction and repulsion, creating a balance, a midpoint, the annulment of the opposition forces, the lobes of the thalamus.

That creative fusion arises from the alignment of forces of different emanations at different distances.

To the find was in a same orbit was added creating a phenomenon by first time manifested. A double energy conjugate in a new dimension creating a polarization interlaced in the lobes of the thalamus without tear apart the one to the other. Rather than the imagination of the ancients only knew because they used it in their resurrection first law, and meditation exercises. Of that polarity is creates the trine force. A fourth force comes into play from the outside, to influence other functions, periods of time and the force of spiritual maturity. While these energies interact within the brain engulfed the entire structure in a Corpus Callosum, or skull, to regulate the rates of energy that would enter his cockpit. Your endocrine body already in development only thus would leave to spend the amount of frequencies with which it's interact will be harmonious. In addition, the thalamus creates your cabin with the same principles which regulates the energy and vibration, frequencies that leak to the quantity and quality of energy that would combine in its Center. Axons and dendrites that seconded it have also their shock sequences, as well as dendrites and their terminals that development to channel those frequencies.

35- Testing study on the Atomic interact.

Based on the quantum theory:

"With the new century is born the new doctrine; on December 14, 1900 suggests Max Planck (1858-1947) the innovative idea to consider radiant emission as a batch process that occurs through elements isolated from power, holders of a certain magnitude. Such element, the quantum, it is proportional to the frequency of the beam, the proportionality factor being a universal constant of nature, the famous constant (h)_ who would later immortalize the name of its discoverer. Thus, the energy is given by the formula.

The lucidity of this thought clarified slammed the enigma of radiation from a black body, immediately explaining the variation of the curve in campaign, whose whims had baffled investigators. Such success was no more than the first exploit of the new theory. "The hypothesis of Planck was hiding the fertile seed of most of the outrageous and wonderful ideas that should transform, until it unrecognizable, the image of the physical world."

"In this dimension further altered by this energy field, is not that it recreates the diversity out of the influence of other substances.

This gives you to be developed after the gland thalamus, a unique quality to become aware of all his attributes. A brain with superior to other than created animals.

Conditions that allowed to get rid of primary energy and are in harmony, with the same creative energy that generated it rise. Share, (open a channel of attraction, and mingle) with the source where I discussed his condition. Communion as you can pure energy that can attract the one for periods of sequences and need to call the other and merge into a single universe of beds consciences than they are." The notion of life before death, emissions and emotions and passions that produced in a newly created awareness, enjoy the first emanations of the cosmos in the universe. What is has of manifest, man and woman contained in a same egg and is reciprocal of its Emanation, in its interior. As radar that returns the same signal that is received, with the ubiquitous reflection of its counterpart that comes together in harmony to create. Per the need and the harmonic scale necessary to interact to the quality of the commodity which had a capacity of attraction; it was the subtlety of energy traveling to that commitment of creation. And in the consciousness of being mystical elevation of entrance for short periods of time in contact with the mystery of divine power, only channel of exaltation to the cosmic, Gods. He only imagines that that gift can be achieved by human beings or that in specific periods of human history has been a knowledge that has not been lost, which is affordable to everyone who could be harmonized with (I), imagine that it must be a gift for seekers of divine wisdom. Something is to which many aspired to own. In this world analog of meeting of forces already created, by evolution of the primary force interact of them needs of follow a pattern of evolution towards the demonstrations already created. Like a scroll or duplicate of the first, but in a dimension, far away from their original Center where features differ in waves of light and refraction. The unbalance of qualities of attraction and repulsion, a need already expressed passion for further projecting to things new and unknown. A new heritage of primary energy that was attractive brought a harmony of its parent. As a mother's womb to give birth a new creature and energy of emotion, he seduced her to follow a pattern of conduct. As that manifests itself in a new flower and its scent, with a superior will not owned by anything, just the being created in the image of the divine energy.

The awakening

Without realizing it, I opened the doors for the mystical development of ancient wisdom, which envelops the aspects that were awakened in me, be asleep for millennia. In my conscience is harmonized with them in a tear of powers, which joined the abstract world, not being with the border. The consciousness in the world of physical expression, a kind of cosmic initiation, a feeling that opened the canal where many Mystics have been touched with this link that lies beyond the human reality. Enter in contact with that sublime force creator that reveals the other side of the creation the portal where crosses to the other side of the light. Of any old school of the mysteries to the initiated this revealed or attracted universal mind rite was saved. The awakening of the subconscious that it covers everything like lightning are switched off and is lost in the most remote corner of the memory. A torrent of energy blocked, of time it, forces rational of the matter I transports to the world of the irrationality where it spiritual is confused with it real and gives a vague notion of them worlds of it internal creation. Since being volume awareness of divine creation and powers which give rise to fluctuations of these routes that cover the development of the universal. Worlds that have been penetrated by many search engines and beings inspired by God our father most high creator, infinite logos of the universe, so we call it because it is universal to created beings and their poor understanding. The inspiration of so many mystics and avatars that were surprised to pass that threshold, entering stage lighting, as well as many that preceded it. Legacies are the reaction of those who create mortals God messenger's God created the universe, the material realms and finally the self, which acquired his large spiritual riches. The biggest secret is that it was his creation and deposit in the greatest treasure, the realm of creation, the realm of the inner light, a universe like the Grand universe in miniature. Where all laws are contained in the divine mind, the lost Word, the word of creation, the Grail (of creation) that contains the true name of God father-mother and all the names of creation, not to be before becoming real. This link lost and blur, camouflaged in the consciousness of the self, where it was recorded for the first time in the thalamus, the womb of God.

That begat itself since the first glow of evolution, where the father-mother - of which his name may not be pronounced is recreated, whereas the light from the heights, the father of silence; the light of the word, radiation of the divine powers towards the Earth-generated auto - impossible to be captured by the human mind - where the light cannot be created by the light - being cannot decrypt the be - vibration which put in motion the first law. Where its duality created the vacuum of the demonstrations, was movement and rest, to the divide grew and grew, concentrating the forces of the opposites in the Center. To be dual, father-mother, and to concentrate its forces in its Center, is the son of movement and rest - the law of triangle that dominates all human manifestation. Appears to be that it is the only one that makes God and realize their self - interacting vanishes from its first essence - but it does not lose its creative quality - love of harmony to the divine as a binge of divinity when creating new dimension. The mind of the creator is the place where it was contained from the very beginning, emanates to be created out of love. In being still slept and buried under thousands of layers of creations and free will, the strength of the freedom to choose, sometimes erratic notions. Human consciousness has diverted the divine connection and disconnected with the Ethereal. It immediate it material is seized of his consciousness, by the lack of harmony with the fourth energy that it says. It is the work of the divine architect on Earth, the cornerstone, opens a channel of attraction toward that force upper, the part that gives that legacy to be, and is part of his way to it. The cornerstone that architects rejected because they did not have real knowledge and were dazzled by the agate nature stone. His intellect was diverted from the primary source, and they lost the orientation to the present.

Secret of secrets, the silence of God

Those first beings- builders of temples, heirs of the mind of Adam, Jesus brothers and followers

This fourth force and the lack of use of intuition, it has buried in the most remote corner of the memory. We have the power to awaken him, but real trails, correct lighting to attract that strength to us in its original condition, have clouded and it has become a dark shadow of our consciousness, the secret that dominate the alchemists, who imaging and enunciate these principles. Every day has been removed by the material ego the same beings, misunderstanding the lack of orientation towards who we truly are. The union of them opposite of power, pure light and darkness, in the Bridal camber, the sacred place that is content inside, the seed of the creation. The confusion of the divine nature with materialistic reactions, to the spoors and drunkenness being must return to light, and soon will be on the right track. It is time to abandon old concepts that do not contribute anything to the spiritual elevation of humans. Wisdom piousness it old, is pure science, like it or not the detractors of such notions of wisdom of the early Mystics. Them were the approach right and left embodied in their legacies the truth so the human is owner of his own free road to the father. It is the divine law and the relationship of the self with the creator is direct and personal, of each being to be in different stages of evolution and understanding of its internal attributes. The contact and exercise that power be aligned with that attribute abandoned for lack of orientation, where we were away so we accept to receive this additional gift of another human being, the created idols as gods.

41-The creation

Statements for the first philosophers assume a Creator God who acted in the matter and the human being by means of the verb. The manifestations that combined heights light with matter formed in the Kingdom of Earth, which is the only one who could conceptualize the phenomena that manifest themselves in the projection towards the Earth and created kingdoms. The main reason for the study is the human being and the way was created and how it works the divine mastery of its laws. Phenomena I describe this system to internalize the divine emanation of God in the labyrinths of the human consciousness and its attributes that include his personal and collective logos. Descend from the spheres of light that encouraged as to matter that connects the Crown with the divine realm. After a stage of manifestation and maturation, the reversal of the process where the energy of the light returns to the Kingdom for their regeneration or purification and enter a dormant stage until their time of the Act, and can flow back a new creature at the time of birth. Here I give an example which describes this inner journey and all that happens when emissions interact with the conscious mind.

My initiation lives the freedom of feeling that the divine energies of creation, a path expression, which occupy its spaces within the functioning of humans and that they can participate in that process. Mature indoor training and make real this spiritual divine gift within us without anyone, this dual process of being outside and God, is interfered with. That was the form in that Jesus and the first settlers interpreted of its relationship with logos. It is clear from the Hebrews, Jews, and the cultures of the region where the Old Testament and not changed history arises. He is mentioned in the ancient texts that were left who took control of the institution to collect the ancient writings, guarded by the followers of Jesus. The stages of education, training to all those who joined the Essenes and other schools of wisdom, for maturing knowledge and through an advancement and preparation provided the opportunity of belonging to the inner circles of brotherhoods. The method was to maintain the purity of the teachings, which were available for all search engines. I went down to the Interior man hell, my own being with all my attributes. I unleashed a soul in decline testing, I saw how as he began to shine his light with property is exceeded and its essence was moving away from the material presence. Inside each day is more illuminating and overshadowed his own being while returning to a State where ceased the advancement, deflected by reasons that marred my understanding. I knew that way and followed him, and returning to the trails they bifurcate to me. Had no idea at first that I would have to continue to learn the gift of the elect, the wisdom of the adepts interior lighting, to which may be guiding my desire to know and learn about. Just another hopeful was being that I was looking for the path, jumping emotions to then defeat; internal inclinations doubt the lack of certainty of what is pursued, it clouded understanding. If is looking for in them legacies intangible, is loses sometimes the trail, although it creates safe. Walking on waters without seeing the veil that sustains the soul, feel the hunger of knowing is lost and wants more, see to her around like everything is crumbling before the futility of the search. Be end route, track and time crystal that is hidden in the trail disappears under your feet and the soul sinks. Return to the beginning is already a routine without return, turns to undertake the path and the mysteries flee hiding between the walls of the time, the ghosts created by deviated from the true sense information, have ended in the same crossroads. Which maze of shadows looming on the search engine and suddenly you see a glimpse of royalty and it raises the soul, you know the facts, it starts the darker path in which humanity has travelled. But not in its material and spiritual attributes, but in his poor soul that since she was born and released his first love to the created distorted itself light, your original divine heritage and the purity of his Kingdom. Was looking for more light than the genre moved away more of its creator, outside of the glow that let it flow, float and move like an inflated balloon of his own being. The ignorance of the true path was diverted by those who introduced a practice away from the original knowledge. Cruelty tyranny led to the deviated from what must have been the rebirth of humanity. Those who had original knowledge not provided for these contrived creations, and nobody could understand the message of the master, as their purpose was that his people not succumbed to the power of the Empire. Those were mystical traditions of the Essenes and Egyptians, Hebrews heirs of the Mystics of India, where the original knowledge was developed. They were only affordable to those tested in their arts. To know the divine essence, it was necessary to pass a preparation before being admitted to the preparation and demonstrate at least a year the wholeness of being worthy.

The legacy emerged in the land division of the kingdoms and the creation of what would be the emanation of the father creator and the universe was working its way in all directions. The essence of creation that travels through a tunnel of light, rises to the cosmic portals where once crossing won't return until his regeneration trained her to inhabit a new body, called the human death; by mystics, transition. This back to life then is reincarnation. Power light emanates to the egg of creation after a period of regeneration, will come to light, his breath to enter the cosmic soul. He is housed in the new being that it will cry to their parents again in a material body. At the first dawn of his awakening, he manages to take control of their new cockpit and start your own domain and spiritual attributes that accompany it in this walk of evolution will join the new matter.

Creation Steps

The return of the power of the father, the embodiment of its light and its being through the union of opposites and their new return to a new creature which will link part of their essence with the new live your light, so that a soul is a reality in this earthly plane. The same law works in the other kingdoms created by the God most high. Except for a single condition that human beings differ in its creation.

Page 11 (the amazing human body) Reader Digest - ISBN-84-88746-23-7 centuries that free human being, a single detail of the creation, the female quality, hormones, estrogen determines the maturation of the sexes or the maturation of their sexual organs. It is an explosion, in the information stored in their laws and the first genuine and reliable detail of how it all began for me. The clitoris of the female takes place in the same area of the penis in males. The brain endocrine system is that stimulates the ovaries to release estrogen. These female hormones are responsible for the maturation of the sexual organs. I took days to meditate in this new phase of creation. In this way in a single body alternate gender in each incarnation, giving for granted that being originated as an entity occupying one body and hence they emanate all the features of the evolution in the alternative form. The perception of a space-turning spiral, climbing and travelling back to establish contact with the cosmic forces, in the first stage of the creation manifested it in salt water, female force emanates first and upper-scale manifested complement the law of creation of the divine energy of the father and the egg of creation was formed. *<u>life became as known for all eternity until the latest effluvium that being can grasp, because he is the carrier of his Kingdom. Being which have been created to give a light of the divine nature, for the ego of our way of control and ignore the divine nature of God in our being, because it has been diverted for other beings that try to preserve this knowledge. They have not had in its form to publicize something that persists and is consistent with universal laws. Those who projected an idea to promote have</u> embarked on systems not to leave this to flow free minds, because they blocked the legacy so that humans remain ignorant.*

Only be affordable to the chosen and prepared to handle this knowledge. With these original spiritual gifts that have managed to preserve is a cache of secret formulas and arcane methods of what should be routine. As life progresses, more, is wrapped in layers of protection what should be natural to humanity as the carrier of that inheritance is the Kingdom in its being. It is the human being, which inherits those functions in his being, should have access to that knowledge and become aware of this gift, the own creator has placed in him. Others are elitist to use systems created to manage version acquired property of the followers who must maintain an age-old tradition that sometimes bordering on fanaticism, because thus was imposed, how to make it known. Long before this period of obscurantism and domination, advances in science and health towered as an advanced branch of humanity although they were involved in an imaginary world of ancient god's creations that have come down to us in books and stories. The contemplation of force disproportionate between men and imaginary gods to give an explanation to what you imagined. His deification of them beings by their qualities, making the sacrifice of his own life, delivering his soul to the God in sacrifice, this was the ritual of them gods and ancient deities. Examples are live daily and not by scholars, but by ordinary beings that do not dominate the sciences. The nature of many human beings that have last by that experience personal and have returned to the life, can give a story of that travel astral resurrection, where not crossed the portal and had the opportunity of return to its same material body. In no way, this trip is a radical change; it is only an experience of the bridge of the conscious and the subconscious. Experiences of which human beings are endowed with these attributes are part of his divine heritage.

Reincarnation

Just see and touch your expression on this flat earth. We did it as a matter of routine in our daily living. There are few who are passionate to find parts that come in the game of creation. Many Mystics went in search of principles or causes of our demonstration on this earthly plane. Their discoveries paid field so that human beings become aware of their own existential reality. Them civilizations that have contributed to this knowledge have been studied, the greater of their contributions to the humanity is has lost, them wars, the destruction of their legacies, the Suppression of its original for impose something fanciful, that benefits, to their creators. What has come down to us is the collection and disclosure of one part of the story; the other was hidden or destroyed by the historic debacles. The other keys were buried under a dark knowledge and only a few have managed to penetrate that wisdom. Many concepts have been swept away from public knowledge, because trying to document these aspects of the universal laws, their creations lose value and set forth doctrines are evicted from their power to create a vacuum in their statements, a vacuum that dull the path of light. A vacuum in its content, which no motive being to elevate his intellect perceives the God who created us. A content filtering human emotions to focus on a figure that existed at one time, so the notion of God is confined to created beings and not the divinity of the father of the creation. It fun or divert them emanations, of the souls that inhabit in the flat physical makes that their emotions of divinity inside are focused in a concept that is not the correct. The degradation of perception diverted to messianic concepts, where perceived an income by describing a message of what is believed god is. They do not feel respect for the principles that proclaim abuse of power, the misery of the poor, not he is kind-hearted. Otherwise, uses their resources extracted from the dishes of the needy to speculate in politic, without fear of offending the God who proclaim, are whitewashed graves that are not unmute for breach of laws. They are the least that they are elevated to the charity and are afraid of God. Education systems do not provide a space to alert the interest of seeking universal, spiritual laws that paid to our existing. We maintain an intelligence which varies in degrees of appreciation per our imagination and preparation and spiritual inheritance that returns in a new soul, a child. Globally, we maintain relations with those who are cut off from us by material and intellectual barriers. There are few who have managed to overcome the fear of persecutions and threats of the statements that were imposed on humans under terms of controls to protect a heritage. Still survives a subconscious culture in this stage of civilization was experienced. Now, with the advancements in communications around the world, everything is within reach of the hand to anybody; the world has shrunk; there are no barriers that stop the knowledge.

Must take advantage of that gap of freedom that is have opened for our advancement is consecutive and real? Those who hold the power of education should provide a system to raise awareness in the educational part of this knowledge.

The original legacies

It was a sprout suddenly, and after a process of evolution where they were developed for millennia the attributes that each of which matured physically and internally. The awareness internal and emotional us trains for accumulate features, that light us of a perception that dominate the universe to which belongs our planet and the humans that it inhabits, that, by ended, are them unique beings aware of part of this cosmic scenario. So far shared without having knowledge that apart from everything that manifests itself in our planet, known or yet unknown. Out of that data is a mirror that must someday come to know. Since the first settlers, we have experienced the most diverse sensations and coined them within ourselves as something bigger than us that we worship because there are many of these emotions of spirituality. By spirituality, the human being doesn't think clear. Assume that is the belief in something divine, of an attribute of God or legacy that we give them higher emotions, to which we have access. An internal gathering, reveals to us something greater than what daily thought, feeling emotions, admire nature, seeing the face of a child, bring us closer to something that is proclaimed as divine, was a face a relic. These are internal reactions of which it has projected one's values that is catalog of spirituality. We try to personally explain what inheritance that grows with life itself, we mature by evolution and application of new sources of imagination to knowledge. Each being contributes, as a separate entity to the great universal mind. The subtlety of these is at a level that we do them in our interior and the daily live, but our vocabulary cannot describe with words those domestic realities. Philosophers, physicists and scientists have expressed regarding this, giving the most different explanations. The causes that fly in the creation are not in agreement in their conceptions, rooted in our minds since we study the concept of our own existence. Scientists are still seeking to demonstrate that quality of the origin of life. Oceans of ink in libraries from the stones up to today's technology where time at different stages of internal events, our imagination and capture it in some means for communication. The science, the genetic, the Astrophysics and others branches know and catalog the knowledge of it cosmic and Atomic and its functions have advanced centuries of information; to catalog the spiritual and physical than comes from cosmic matter. The magic of the mind over matter, the desire to create free of controls, gives us the freedom to wisdom divine to enter in control of certain direct powers of imagination, join one point with another and reflect its essence in all directions is to create a new underworld; It was not in the mind of the self, but it is possible in the divine mind has no limitations. Just realize comes in relation to the excitement of capturing the attributes that the divine mind has access. With this simple step is that he reveals the small door of the universe in the soul to be seeking uniformity of all laws. Perhaps sixty years ago, would have been a great job overcoming these barriers. Since the Moon is stepped for the first time, human took a leap in its development that has not stopped. We overcome the concepts that were accumulated for millennia by the best minds that we provided basic knowledge to educate ourselves and follow the accumulation of data in our respective inner worlds. Colleges and universities still use these legacies of the antique as a base of those phenomena that we bequeathed. The most significant change was fun mental stagnation of religion with science. Put life in new perspective, overcome concepts accepted as truths. Knowledge must be a tribute to these mentalities. The world is witnessing that a new era of knowledge has aroused and maturation of human intelligence based on concepts that challenge the most established schools, although still need to mature a world of concepts, not to miss that link, that door which gives us the freedom to know our past. The large mathematical and scientific that molded them bases in it cognitive, as a universe abstract of know and the projection of the thinking human towards an extension of the power creative of what exists and it by create. Archimedes was one of those teachers outnumbered few years before Jesus, its treaties are being rescued by scholars that, although they have given to the originals are they busy to give mankind hope, in 2005 began the recovery of data from your C treated, disappeared from the library of Constantinople.

The code of Archimedes manuscript that could have changed the course of history of science-Reviel Netz-William Noel (ISBN - 978-950-04-2926-9.

To who you should not this be discovered, were released by the Vatican, this means they are protected since disappearances, things are changing thanks to historic pressures on them.

The great defenders of truth, thousands of names were contained in the libraries of Alexandria, that nobody knows his destination, that of Constantinople which was treated without mercy. World Centre that will provide the humanity of the ancient wisdom and advancement without so many obstacles that must be overcome. Laboratories and developed minds of humanity waiting for that great outcome of the hidden history, for being drink to their original sources are ready. So, the ignorance that has been maintained for centuries can be overcome.

The way out of materials concepts that we alleged a universality of all attributes of matter and energy in the universe. To that goal intellectual forces of the universal youth must be addressed to achieve together real concepts, for lasting world peace. The day our planet cannot accommodate more beings, our youth should have ready the way to colonize the universe. Gaspar (Edwin) Pagan- July 7, 2010)

51-Evidence of evolution

The Crown Keter, the thalamus in the center of the human Brain. See figure in page no 22

Esther Roson Gomez of this doctor and scientific contribution to the science of the future: a simple way gave a comprehensible input, original human processes that still exist in our original system. In the scientific study published by Dr. Esther Rosón Gomez in 2005, you can see perfectly this center of the body. Each being has an individual personality, which is linked to the functions of your thalamus, associated with endocrine and hormonal system. It is the field of action, the barrier of time space in the human body, where regulating characteristics reflecting the accumulation of the personal evolution.

Recorded in the neurons of communication in each area that is affected by the actions of the physical body. The regulation of emissions which interact with the energy of matter formed in beings from the beginning. Emotional, spiritual correlation is also per the ability with which the imagination, emotions to communicate or interact with cosmic vibrations. I mention cosmic vibrations because we are receivers of energy traveling through space towards our worlds and bodies. Its vibration is manifest in atoms, molecules and materials with qualities that we endow of a variety of reactions different. They make us aware of the forces unleashed in our interior. We create with these forces. We are breeding grounds of the divine mind. For the moment and in the century XXl, knowledge barriers are rising. Daily new steps occur in the discoveries that give you a real idea of what is perceived and is believed. There is more certainty than doubt of internal processes, announced by science. Soon won't need this effort, so that the minds that have evolved and left behind this time of fanatical obscurantism emerge as the new pillars of the development for the future. I hope that, with humility, they will approach the future of humanity. The minds young must know this legacy that is maintained to stripe by them systems created. In Visual imagery, this gland that retain the emotions caused by reaction to Visual effects, effects of the spectra of white light that is phosphorylated in their processes, vibrates per the reaction to what is perceived, causing that they wrapped the range of emotions ranging all the manifestations that are attributed to the human being. Laughter, fear, rejection, acceptance, love, admiration and all the attributes of human expressions, are an attribute of the different arches of internal harmonies. I say arches because it is a jump from sparks from one neuron to another, causing these effects of alignment and balancing of creating internal waves, which affect our endocrine system. There are even studies where is attributed to this part of the brain, an area as old as thalamus, fear or the acquired notion of protection against the dangers. These ranges of reactions we experience interact between the cosmic mind and gradation that human beings can experience their relationship space time in the cockpit of the mental physical being. It is a natural function that was developed from the beginning of time; it enables us to operate all the defenses of protection and escape. The unknown, the intuition we have dormant within us, a sense which alerts us as a spark that it covers everything and gives us an accurate notion and sense at times early. As every human being who seeks to unravel a concept or explain processes, go to the appropriate sources. Therefore, I should go directly to the teachings that encompass the notion that advances the idea without having to produce it. That is a routine of those who seek to make something known, because no one has all the sources of knowledge. Summary of the discoveries of Dr. Esther Rosón Gómez published in 2005. Resting and action potential:

1. Location of rest: in this situation of repose, the membrane of the neuron is polarized so it is 90 Milli volts more negative on the inside than on the outside. The potential of sleep is of-90 mV, due to the concentration of sodium and potassium. Note: "nanotechnology may include deep in the system of energies that work in harmony with these membranes formed to filter and decompose the radiations of the creative thing. We know only a small universe of them, since it is a new field and I'm not scientific I mention it logically must exist for what I imagine several discoveries, which does not have access or otherwise minds prepared to pay to scientific knowledge. 2. Depolarization: is open those channels of sodium and the sodium enter in the interior of the cell. 3 Repolarization: sodium channels close, potassium channels open and potassium exits cell to replenish the negativity. 4. sodium/potassium pump: expels three SODES, for every, two potassium's. 5. Recovery of cerebral hemispheres relaxation potential. There are two hemispheres, one law and another left, which

are separated by the interministerial cleft. The cerebral hemispheres are covered by gray matter and cerebral cortex. * Note: "these materials is accumulated in the cockpit of the brain by eons of time, during the reaction to them powers that caused their accumulation and the need of material to them neurons processed that matter." Where were the basic materials for attraction of the compounds that gave base to the creation and were distributed through the channels of the material body?

Following the laws of grouping materials, necessary for survival, channeling of energies that were creating empty attraction forces to complement the work imposed on them to achieve harmony in their creation." The cerebral cortex has Gyri and sulci that increase the cortical surface.

Cerebral cortex

Voluntary movements are made. The sensations are made aware. Information is stored. Prepares the psychic functions, located in the Recental ring road, in the frontal lobe. The movement of the voluntary muscles of the contralateral hemi body origin orders originates. The fibers that are originate via pyramid form. It is in front of the motor area.

This area program movement: It has many connections with grooved nuclei and thalamus that will act as centers of connection. Note: "based on these observations is shows that exist, a hotbed of reactions of energy, creating the environment for those processes are possible." This aggregation processes have a maturity of millennia and the strength that led them to this maturation; It must be a so subtle that only the divine light of creation, can act in these processes. Its qualities are maintained as in the beginning or with slight variations. They must have suffered millennia of variations in material changes and behavior modifications that were refined processes that possibly even science is trying to document. The energies that have created this scale of conductive fibers for the wide variety of energies that circulate in this skein of nerves and neurons, which are its complexity, give the trashing of the imagination a progressive process of overcoming and create channels to modify internal processes. Dual being that energy enters the internal process and ramifies through their channels created by evolution".

Area oculo cefalogiria is in the lobe front contralateral. Voluntary movements of the eyes and head are given. Area for the calculation, the recognition of the body schema, recognition of touch and reading. The region is located at the end of the Silvio cisure. Area of language is almost always on the left hemisphere. Divided into several areas: Broca's Area - level front. Wernicke's area - and in front of the occipital - temporal level. Failure mechanism can lead to: aphasia: failure of comprehension or expression of language mechanisms. Dysarthria: error in the motor mechanism of speech organs. Note: "these notions of flaws in so-called normal functions should in any way be affected by the course of a stream of energy. "Yours interact with them, materials accumulated and that was deflected by a malformation of any source deflected of his runway and avoided the function correct to which was intended." LIMBIC system: Region controlling emotions, motivations, emotional behavior. It is wrapping to the Corpus Callosum, the area of the frontal, parietal, and temporal lobe.

* Note: "with the abandonment of the power of control of the will, many functions have been lost. The maturation of resistance, the abandonment of overcomes the will of challenges. The correct direction of energy cooperation with these reactions weakened

the structures of overcoming barriers. The maturation of new networks of deviated from the uncontrolled energy attacks to overcome the pain, anxiety and frustration. The lack of appreciation casts a particular form of neglect of the memory and the attitude of abandoning the will to fight and to recover these areas."

The memory area there is a certain area. It has been given much importance to the hippocampus. Grouping of nuclei- It is next to the third ventricle. Get sensitive information. Involved in motor control- It maintains the alert.

 * Note: "memory of the brain does not have a field specific, because each cell of the body is a brain power and has its own personality in their habitat and dominates a wide range of energies. It brings its development and accumulation of information sharing with the swarm of neurons and nerves through the networks of communication vary, as it does not have a fixed pattern of behavior and be adapted to the needs created by the same system." Striated nuclei: consisting of the caudate nucleus, putamen and pale. Make control engine.

Note: The colors of substances give an idea of the powers of energy that must absorb or filter, turn into a strainer of vibrations that must be regulated.

Contribute to a process accumulations, depends on the specific role of the qualities that must be supplemented as well as any other matter that accompanied the evolution of every created thing. Hypothalamus, region formed by grey matter. It is next to the third ventricle, ahead of the thalamus. It is the main center of vegetative. Note: your function is it of keep a store of matter, be the supplier of them components necessary for is manage them materials that will be then used in the system that are accumulated to be ducts of energy creator. Brain stem brain stem is formed by: 1. Mesencephalon 2 extrusions. 3 bulbs. Midbrain, substantia nigra, where necessary dopamine is produced in the motion control. This area works in coordination with the basal ganglia. MOC- third cranial nerve center. Centers responsible for the State of consciousness and the rhythm of waking. Extrusion core - sixth cranial nerve – Mo E and the MOC. It is the output of the facial and trigeminal nuclei bulb area where it crosses the pyramidal respiratory center, output of the hypoglossal vague nerves. Vestibular balance centers.

Author: Esther Roson Gomez year 2005

The pump sodium potassium ATP-gesture, tri-phosphate-is a protein transmembrane that acts as a conveyor of Exchange antiport-transfer simultaneous from two different directions-that hydrolyzes ATP. One is AT. Transport P type happens, i.e., undergoes reversible phosphorylation during the transport process. It consists of two subunits, Alpha and beta, which form a tetramer integrated in the membrane. The alpha subunit is composed of eight transmembrane segments and it is the center of union of ATP, which is in the Cytosolic membrane side. Also, has two centers of union bilateral, extracellular and three centers of union to the intracellular, that is are accessible for them ion in function, of if the protein is fosforilzada. The beta subunit contains a single helical transmembrane region and does not seem to be essential for the transport or for the activity. Note: The powers thus described are part of the divine energy that began to create diverse matter in being. This pump is an electrogenic protein; since it pumped three positively charged toward the outside of the cell ions and introduces two positive ions in the cell interior. This involves establishing a net energy through the membrane, which contributes to generate power between the inside and the outside of the cell, since the outside of the cell is positively charged with respect to the interior. This direct electrogenic in the cell effect is minimal, since only contributes to 10 percent of total electric potential of the cell membrane. However, almost all the rest of the potential derives indirectly of the action of the pump of sodium potassium, and is must in its greater part to the potential of rest for the potassium. Motion and rest, an equation of the notion of divinity note: The creation of membranes of different thickness is due to the ability to filter and graduate energies so that their role is appropriate in the creation of new substances and connections in the scheme of construction of matter in the body. It is the most important role in the regulation of activities to supplement the body's control of vapor that penetrates precisely this factor allows the intervention of the creative will of imagination, enter in control of part of that energy to establish a communication or frequency that energy capacity and enjoy a high feeling of divine consciousness. Harmonize them fluctuations internal of the matter that we consist, tune your structure conductive, to open a channel between the conscious and the subconscious, the border of be with the not be or vice versa. It is essential that our being perceived that feeling of inner harmony. V emanations of the original Kingdom intelligence, wisdom and knowledge itself will create wisdom that concentrates all the things, the intelligence to stop ideas by Association, emanation, and induction. Grace, love and mercy, attributes of being inherited from the divine mind, the power of the trial that always accompanies us, perseverance until the final victory, the grandeur and Majesty. The Foundation of all the forces that have been activated since the first glow of that will, finally, the Kingdom where dwells the God of Creation and brings together everything in its manifestations, consciousness. Note: The will of projection, is the primary source from which everything emanates, if it is the will of the creative force and its course was the creation, as it was initially that it must have covered a moment in the divine mind represented in Kabbalah. Concept accepted by the mystics that diagram the divine order. To transfer a vocabulary not spoken that keeps the keys to the creation, that being can declare for the future without loss of content. Correct teachings, the same that inherit from the caves and the stones of them ancient. The resurrection recreates the spiritual legacy of the ancestors and was carried out in caves where meditated for periods of time, avoiding disturbance. The reaction between understanding, intellect and wisdom, create the dimension of space and time (the greatness and the force of attraction) away from other elements, distracting internal concentration. God created a vehicle to express their essence divine. The human being is the heir of that Kingdom. Its creation was not spontaneous; it was through many stages of evolution until it matured an expression that it encompassed everything. He united of up needed of the United in them levels of maturation e exchange of energies that interact and attract of the source primary by adaptation, the attributes that must manifest. For that we are created and function as the emanation of a higher consciousness, you need a lower awareness to occur that it is and take life at a level that he is not and thus perform his own existence, while it manifests itself and recreates itself. Humans did not appear of time running through the land already facts completely. The cycle of life is repeats incarnation after incarnation, in all those kingdoms that is the law existential. By that time tunnel space is that travel the creative forces to comply with the laws of evolution, and by that same tunnel return forces to regenerate itself in the primary source. What manifested itself first was by laws of creation, the other following is the evolution of that creation, arise from the creative mind where human understanding possible capture. Mentioned and linked as tunnel that acts as a channel of communication between two sources because energy travels, per its capacity of attraction and repulsion, so that this energy is not contaminated or attracted to others of lesser quality should be subject to laws of manifestation and regulated by their attributes wiggling and aggregation. The reincarnation for centuries endowed human beings of power supplies and the free Act to mature new expressions that later, returning to the original source, are returned to the great divine mind. By cycles, these same powers would return to a new body to follow a growth to exceed its previous state or undergo degradation if the process fails. The power that created us brings us back to its parent company to equip ourselves with new energies and we vomit back to earth so that we prepare ourselves on our way to purification. It is the law that we imposed, to evolve. The demonstrations since the beginning of time, the physical creation and material in nature, and spiritual, with the cosmic attributes of the soul, where the own cosmic creator is recreated in the manifestations of those minds that are harmonized with their spiritual laws. The purest emotions of being that elevates their spiritual knowledge and transcends the material ego, it raises its essence to the cosmos, or what we call God of our consciences. The raise and grow that energy and reverse it to the God of creation enables us to be bearers of spiritual light.

Thousands of thinkers that we admire in the schools by their wisdom, the significance of their knowledge takes us to the recesses more inmates of our own nature. What comes from the divine mind through the human mind is and has been processed through this center of creation, where the internal processes through reside be everything created in all branches of knowledge, which we contemplate in the history of mankind has manifested itself under the same universal laws. Just be aware and has processed with knowledge all that exists and created by man. It has also taken understanding of what surrounds the cosmos in the varied motivations of nature. As I said before, we are at the gates of knowledge that long ago had surprised the common man. The last creation, triangle, create the basis for the generation of new beings and their powers of procreation. In any stage, what was a manifestation trine perfect is divided so that the forces are harmonization again, when all were one. So, the end manifests the realm of divinity. There is hope in the preaching of Jesus that being must be evolution found the road to return to the divine nature which created it, what he called, and the eternal life. It is a challenge and is clear that creation took place in this earthly plane. By deduction, this is the level of containment where conditions for Holiness, or the Holy Spirit announcing Jesus gathered. If the divine plan of creating this plane by any need of expression emerged at this level to develop realms which we see today and what can be captured in the future. By Jesus is the stages of initiation school's large knowledge update case, carried a simple projection for minds that was heading must conclude that is a plan of her own creation. We are the winners of realize us that we have the only notion of the divinity within us and we must look at this opportunity with devotion. Human heredity and evolution human beings do not descend the monkey or primates, as it has been trying for years.

Are great theories that are have developed by great minds of the humanity. As other knowledge, has filled a void in the human experiences nature in its material bases projections is the only way in which the functioning of the human being has focused. Science has been responsible for rectifying those errors of assessment, which, although they are not far removed from many realities do not fill the gap by complex. If seek in the annals of the antiquity, existed a knowledge, although outside metaphysical or esoteric, of the creation; It was in his time a product of human intelligence that was closer to what happened: the creation. In coming times is will shed more light on these aspects, and possible the last link is understood. It is my goal to focus the creation of beings as something akin to the superior forces of subtle qualities that graduate communication with its creator and is related with the attributes that generated it. The designation of names or concepts is a way of giving a space where the concepts together to a universal understanding of what motivates human beings to overcome and continue to exist. If someday subverts is that internal legacy which is common to all would be the end of an existential harmony. There are no divergences in realms that do not contain our notion of something bigger. Other creatures were on par than the human being. The human being is developed, is saw wrapped in them same struggles of survival, in overcome the ferocity of them others, with wit top to them. The fact that their skeletons are mixed in the same environments, and were similar in structure is not a test that one out the heir to another; they simply shared the same atmosphere and the struggle to impose on others. The fact that its characteristics are like human beings not graduated the lineage of other beings to equate to the human being. Reincarnation, endow the man of that special feature, because what returns to the body are the divine attributes of his soul personality, which would not be present in any animal. What has been called soul is a specific function within the Web of emotions that mature within our being. The soul is the quality within them subtle energy that is stored and include the all, in us be emotional inside. An area that collects that vibratory rate and stores with all the emotional intangible attributes that evolve and mature for the period of existence in the Earth plane. A quality that we handle and project as a function of gratitude to the God of our conscience, at the same time separates us from another species. We are the conscience of what they call God and all the attributes of the Godhead operating in greater or lesser extent within our being. The difference is that we realize and mature them voluntarily. All of this coupled with the variety of physical features that make a huge difference internally and physically. A world of demonstrations that we must understand before declaring that we descend from the other animals, whatever separates us. We are a demonstration that follows patterns and laws different. Studies show that we share communion with features of the creation initially, aggressiveness, idealization, concepts, innate intelligence makes us superior, realize the imagination, the conscience of the things. By the knowledge of centuries that accumulates in the memory of the evolution that has catalogued (AKASHIC) files by the mystics at all levels global, distinguished humans from other creatures that exist. Per my best understanding, this is a source in the cosmos where are recorded all the knowledge that ripen our personal soul and consist of knowledge, emotions, personality data. The number of atoms formed our physical being, the exact knowledge of every attribute of manifestation in each volume or cell of our physical and spiritual bodies. Regarding the human being is refers and in its counterpart, the source of where came or emanate maintains a file of all them creations that have issued of its own be. On the other hand, the strands of DNA (the genetic components of each cell) in human beings would have to be the same as the animal, but in that same direction DNA testing vary by 2% of the Primate, who intends to link our evolution. I must confess that I am not a scientist or have knowledge of genetics. If I make a mistake, I hope that emerge a prepared mind and recognized in these ideas to establish a clearer understanding. It is possible that it is already listed. If it were not so, the scientific interest for the proper investigation should wake up. I can imagine that in a matter of books there must be many authors and theories based on this same subject. I don't want to confuse the reader with theories, which do not contribute to something simple to understand. At the time that I started writing I realized that the way in which in the books focus this subject, my perception of these themes, which were far from what the common being can internalize. Meditating on that I've tried to simplify the issues addressed, although it is difficult to communicate something that is so easy for us, unravel or rescue of the stages where he has been buried by concepts imposed by thousands of years. The minds calloused by those same concepts, have a challenge of organizing and accept new reasoning, this not is nothing new.

64-The creation in the salt oceans

My forms of express them knowledge in this book, is a story of mystical, alchemical, metaphysical, since I personally attributed this knowledge to them minds more large, from the ancient schools of them mysteries, philosophers, medical and mystical, including to the larger mystical scientific that has existed in his time; Jesus. The master that treatment, directs the course of his people to freedom, physical and spiritual in its time. In further stages of evolution, have been brilliant minds and mystics who have endowed the humanity of higher education, for the advancement of the least prepared. My way of expressing this tale of experiences through my daily experience is looking for the relationships between the real paths to the truth that dwells within us, which reproduces reincarnation after reincarnation as a portrait of the soul. This method of Gnostic schools where each disciple is prepared to travel to the ends of your inner being, descending into its own Hells and mature spirituality to understand God and human suffering when degenerate his being. The Crown Paradise Escondido

Be it through its evolution

The energies of the combination of sodium, potassium and other chemical components entering the functions of creation, interacting in the Centre of the thalamus, in our central nervous system, endocrine system, thalamus, pineal, hypothalamus, Hippocampus, secondary communication and emotional glands and thyroid, neurons and dendrites, have their origin in the beginning of the creation. They possess their own DNA strand that has changed since its development in the beginning of the creation; simply have been transformed and expanded as a mirror or a source or membrane layer. They produce the energy where primary cells derived their structures, since they depend on these substances for their survival, because they are the source for producing proteins and their derivatives and aggregates. The first cells could not survive without them, and the structure of which resulted in a concrete being originated. As well as planets and bodies in their orbits have a belt or mantle of energies interact with matter forming them. Thus, the human body and its sensitive parts are surrounded by the mantle or aura of energies that correspond to the qualities of the materials from which they are formed.

Very important to define, if this is an external irradiation of bodies, or an approximation of attraction that is present to support reactions that are about to happen. The laws of attraction have that quality as in other realms, used to attract the elements that complement them. This field is the aura immediately to the creature and can be observed by many media where equates to the seven colors of the refractions of light, pure and white, creating original, projection of being not being, the beginning of everything. They recreate an individual magnetism that is the sum of its composition with the spectrum where it is said, the matter is what defines it. The characteristics of the first cells in the salt water have your environment consequently that all forms that were developed initially had a common environment. At the same time, their bodies were developing, evolving into what other creatures were emerging from this primary form. As result of that principle common where all emerges, many of these creatures spawn the eggs, or expel their seminal fluids and waste in the same environment, and being heirs them some of them others, eggs could interact in their primary functions of evolution, attracting or rejecting qualities. At some point, must be creatures of all shapes and sizes, which interact together, by genetic selection survived the most capable? Today, the behavior of sperm to fertilize the egg resembles the conditions that had to be overcome in the water at that time as old to move to achieve its goal. The same conditions are present in the reproduction within the self and many of the realms of creation. The human body is a replica of that process which occurred in salt water, and was encapsulated in what would be the evolution of species, given that quality to the human being. The endocrine system was first developed by the need of primary matter that was common in salt water and which was now in contact before immigrating to land.

They were forced to emigrate in search of the components necessary to manufacture the proteins and substances necessary for survival, thus began the emigration of salt the ground water and land to the saltwater into a period of adjustment. Of the same form its parts external is adapting to them new challenges, where arise them tips and others organs that is modified to comply with them demands of adaptation. To emerge from the seas and salt lakes, the thalamus system began to process sodium and potassium directly of the emanations of the Sun, another key to life. In that process, I think that a skeleton to give support to their sensitive parts and adjust its ability of emigration. That created the diversity of species and the formation of the self. In the beginning, them forms primitive, besides create substances as them proteins and others added primary, sucked sediments of the Fund that is accumulated by the actions of the evolution and them phenomena climate. This endowed them with materials of different compositions to develop new areas of expression. You interact with the same primitive forms, which began to develop materials that are added, and per the environment where they are located. Not all forms that emerged had access to this stuff. The brain that developed acquired unique conditions and I regrouped around specific subjects. Today they are that channeled their heritage of antiquity. If this will change, the evolution would take another turn. So, this can be understood by the less prepared is logical disease be given by any imbalance of the natural conditions of how it was created.

Physicians who provide healing of illnesses, simply looking for some form of knowledge to restore the material being of any substance, which stopped producing in the body, or decreased its level of stability. Everyone is their own world and their attributes are the sum of its history of evolution and the elements that have accessed. Both in their diet where ads new matter for the formation of what will be its structure in the process of advancement.

Note: -in my research I found the discoveries of scientific information twice Prize Nobel Prize in biological sciences from the water.

68- Doctor Linus Pauling

In his studies of biology as the science of water has been discovered that each 1 (liter) of sea water is composed a soup marina containing; 965 cc water, nucleic acids, DNA, essential amino acids, proteins, fats, vitamins, minerals; (complete a total of 118 elements of the periodic table. In addition to phytoplankton, zooplankton-krill - Omega-3 eggs fish, chains of carbon, particulate larvae. All of this is related to the origins of cell life and to my surprise, declares that seawater is complete nutrients from nature. He mentions that add 3 parts of fresh water to the sea Gets a solution like human plasma. Note: - This shows that the human body has the same elements as the original salt water, with a slight variation that I look at your comparison. It means that we have a copy of our material body in seawater. All its components are related to our original physical structure. Indeed, fits imagine that of that arise in the beginning and the adaptations current, the body has varied some of their features, by their changes of environment. Already we are not exactly what we were in the original arise. Some features have varied in composition and use. The variation of human plasma used for testing is the current which merged with fresh water to recreate the variation of the emigration of be of saltwater into the land where we consume processed salt and fresh water. This added to the other evidence proves my statements that human beings originated in salt water. The Gnostic knowledge are the only internal source to harmonize our being rational with mystic and divine, attributes that dwell inside by spiritual inheritance to the maturation of our soul and the conscience. The intuition of the God of our consciousness lifts us to a perception of States of exaltation to meet that divine harmony. To migrate to the land, he faced a period of adjustment and interaction with new direct emanations and the subtlety of the vibrations which would have to adapt. The phenomenon that was exposed was direct. In the variation causes begin to interact and the needs which would have to overcome was very different. Thus, new expressions that need to survive and the causes. About the life shared in the water salt, a myriad of species developed a system complex to survive. They moved from sex to continue playback of their species and achieve this persists. The heirs are the cow fish, Arvid and bivalve mollusks. I read somewhere that adult Hyena also changes of sex, there are different methods of reproduction, of course, our thalamus gland lived in the same environment and evolves with superior features to other species, but similar in form created, endowed her the sixth cluster of attributes about the species: a dual, trine, and androgynous egg at the same time. In them human, the sexes are manifest one at the same time, the other sex is present, in form sleeping. Possibly, in another incarnation the other can dominate the stage of creation. As I explained previously female hormones determine the sexes when mature, the area where manifests the clitoris in females is the same area where the penis in male's manifests. Attributes are contained in the expression of the brain area, and for me the thalamus, which is the gland that defined him and play for the first time. There is no way to declare or prove that the human being was created with similar characteristics to other beings that originated in the beginning. There are so many stories in various ancient civilizations, which focus on the principle of human beings as an androgynous egg. It is possible that your body would share an equal to the other evolution and then split into male and female. Even today these phenomena that seem us rare are numerous in nature, something natural in the continuous evolution. In the part human is a repeated case of androgyny that even has been scientifically catalogued and is reason of study, but within a same body. In a version of selections from Reader Digest (page 194 - the wonderful body) - ISBN-84-88746-237 - I found this statement with a maze of imagination. Female hormones are responsible for the maturation of the sexual organs and manifest themselves in the clitoris. In the female clitoris of embryos is the part where the male penis is developed. In development is the same where a common organ is developed to express the two conditions in the human body and they manifest themselves in different parts of the incarnation of a sex which determines which is which defines the manifestation of DNA in the future. This is the areas of excitation of the clitoris in women and the foreskin in the male. This generates a scientific explanation of how the sexes do not require a different body to express the dominant sex. The laws of creation God made perfect things. Being must marvel at the created realms because in them is given to know the divinity of your creation, I'm personally - and thanks to that sublime being, lighting that has given me this day. In the realm of human beings that were developed in different areas of the planet inherited physical characteristics of their way of surviving conditions of nature itself, which had to face for their subsistence. The variety in the diet of each other, made the difference in their physical characteristics. While one of the breeds used to hunt, there were others who were not under this same influence or separated by arise in different areas. To mention this detail, I assume that the purity of their feeding trained them with pure qualities of life and its development was most advanced in terms of his physical form and inner development. Another detail is that, to the storms and those different phenomena of the nature, the forms original is dispersed in the water by the entire planet due to those same phenomena. If we look at our American continent many plants and animal beings and other species have resulted from these natural phenomena. Your brain is adapted to the evolution of materials, as fruits and herbs, the water pure that is consumed possibly to the fishing and thus the development of them features cell. One very different to those that only feed on animals and their DNA was mixed internally with the animal, acquiring physical characteristics of the meat they ate, as their behavior to be linked to wild beasts and animals which had to dominate by force, their attitudes were of this nature. The absence of a rationality that made them overcome the ferocity of the animals endowed them with such attitudes and mental and physical development. On the other hand, there were scattered races which adopted different phenomena of development and were more interested in nature, their behavior and the ways in which this is expressed. Their emotional attitudes were maturing otherwise and his behavior was friendlier between yours. In other words, their behavior was more emotional. Somehow, they adopted ways of life more organized and respect by the nature itself. Natural phenomena attracted them and sought an explanation for everything that happened to her around. At some point of the evolution the nomadic tribes of Europe came together for some reason either wrestling or submission or by dependence on each other, then somehow reasoned that joining forces it was easier to survive. Some were ardent hunters; others far were victims of another or the same wild animals. Another reason had to be phenomena that occurred before in the Earth which was exposed, more natural phenomena not finished with all the creatures that evolved. Apparently, there were cataclysms, created by cosmic phenomena or phenomenon that somehow lock the emission of energy from outer space, for a period and the creatures that did not achieve development, in their endocrine systems, which enables them to handle this crisis, succumbed. As cataclysm blockade of emanations from the Sun, the interference of planetary phenomenon, which opened ozone in any region of the planet or the immersion of the Earth in a gravitational field of an asteroid, I mean our environment created a change in terrestrial gravitation, or affection with rare material. Causes of variation in behavior development of human race. The other kingdoms of creation are influenced by the energies that we process for the characteristics that gave rise to the phenomenon of life arose. By reasoning that any external phenomenon that change, of the kind, which alter the laws it obeys our development, is physical and emotional, it obeys these same laws. Imagine only isolated physical phenomena can give purpose to the total life would be an error of calculation. Attributing to the violence of the cosmic laws only this behavior, verges on the little imagination of our creation and evolution as if this is materially like the existence. The beings created by the primary emanation obey universal, divine laws about which are the emanation of intelligence that we carry within ourselves.

73- Effects of ionization in saltwater

Theurgy magic and metaphysical 2012

The mystery of creation occurring only on planet Earth to the present we know. The Kabbalah of creation: for my little knowledge of the Kabbalistic principles to which I had access detaches the notions of Kabbalah are applied directly to the creation of the human being. It is a way of giving meaning to those who referred to and is revealed from the knowledge of what is experienced and perceived. They are statements of as it is attributed to a God or supernatural force, the influence it was created as a being. Is our internal process of creation a unique in the universe intelligence? Possible races of other planets are interested in our planet; it would be by the unique notion of intelligent life on it did. We are faced with this question that the principles enunciated in the labyrinths of the cabal should work created in the universe, without excluding extraterrestrial life. A condition that emerges from the cabalistic statements is that forces emanating from the creation are specific and are due to cosmic causes that so far are accommodated to human life as it is known up to the present.

If arose creatures with a mode different of interact with the matter universal would have that design a system cabalistic that is express on all the laws that is manifest in the universe more all them kingdoms of the creation, of that form harmonize the knowledge with the reality existential. If in the infinite universe there are other spheres of underworld or areas where the galaxies are a prototype alien to our life, where another kind of demonstration system is developed. That the existence physical not be required, that only is a flow of attributes infinite or essence of manifestation. I refer to the realm of cosmic consciousness that it covers everything, where the refraction of primary matter arises.

Effects of ionization of sodium and potassium

The functions of the thalamus gland

It is possible that intelligent life on Earth is unique in the universe and if any intelligence outside of our Galaxy will be close to our planet, it would be in search of the frequencies that are conjugated from outer space and accumulate in our planet.

Unique condition

Being a distance only the Sun and the conjugation of elements emanating from the different sources of energies into what makes up the planet Earth, he has been adding since your ablution first matter, is a unique condition, and we own it. Our imagination is raised to be prepared for actions of races from other worlds, it is logical to think that it can be higher intelligences, their lives being very different in molecular structure to us.

Centuries before our environment has been invaded by beings from other dimensions, we are a unique status that derives from the energies of sodium and potassium and elements such as oxygen, carbon, and the elements in salt water, which is nothing more than the concentration of the sun itself in the water of the oceans that surround us.

In addition to all the elements that have been concentrated since the emergence of the universe in relation to materials of the earth itself and its fluid Ethereal as the air, the energy circling us, while we are exposed to the powers of other bodies that paid with their unique conditions, the lifestyle and the phenomena that manifest themselves to us produced by outside forces. The function of the first membrane (thalamus) of creation that gave rise to the human being is one of the main objectives of anyone who wants to copy our source of life, the origin of our race. If someone would like to duplicate it life in the space outside for originate processes of creation should follow the examples of the divine creation. How did it begin? _ What active principle that these elements are organized and start this type of reaction? _ Energy materialized and met the qualities that gave start to the first cell? _ I analyze this in relation to the human being. I establish energy creator interact in unison with the different forms of life that have emerged. It was a spread of some source that arose suddenly and generated these reactions, initially by disorganization. If who would not give an idea of how step, if not the intelligence of God in the created being. There should have arisen in grades ever higher. Their qualities had been vibratory, by its concentration in the different areas of the planet, gave place to the life that arises was in a principle related with another. Begin you interact with various materials, depending on the area where manifested itself, reacted this energy to the attributes of the materials that were added, creating the diversity. This was determined by the frequencies of energy that attracted and combined in their structures. Different temperatures cooperated with the generation of forms. Speaking of temperatures, we find another link where the earth itself, from her, as a mess of nature, brought their gases and primary materials to help the delivery of energy travelling to fertilize it. In the features that block eventually emerges as a human being, they adapted and reacted materials of different characteristics, compounds or atomic variation. The electronegativity is an explanation sensible reactions that happen in the matter, per se, is the key to our spiritual advancement. In fact, the subtle energy that excites us and enables us to realize energy of divinity, the beauty created in all realms, is direct communion with the light of the divine Kingdom that created us to subvert these energies in feelings or real images. Although in each appreciation is meet all the features and qualities of definition, the minds scientific seek accommodating them words exact without leave an opening to the speculation. The saga is that, if one tests something, we realized that someone later reveals that same observation with data developed on science. Salt water was concentrated by the Sun's rays that must have been weaker and that not fully penetrated the layers of dense water was freezing or turbidity. In addition, the atmosphere was denser and made difficult the penetration of light, which was progressively growing within a framework of reactions of the solar system. It is possible that the attraction of internal atoms and their spaces create perfect composite material for this approach, that the sodium and potassium contents in saltwater would serve as the primary elements. From that moment of creation that phenomenon has not ceased. _ What emanation creates this first manifestation on the Earth? _ What record still interacting so that attraction continues as a law only of the creation? _ What qualities have been added for millennia to original attributes that awakened life? Neutrinos, _ what function exercise in the spaces and spectra of matter? A blanket of subtle energies, with known spaces, where travels to corners of the universe. Where it is required its complement, there is mixed and interacts with similar qualities matter. Specific functions of the mind, imagination, intuition, memory files, emotions, which accumulate and frequencies may be attracted to the emptiness of the mind that is activated to collect the former legacy, necessary to complete a function or retrieve a fact. The mind of being harmonizes with this natural function, in some minds that meet that information to their advancement and benefit. Particle or combination of particles reflected one against the other, as a mirror of refraction of light, and have the darkness of clear light that invests in the speed of creating - see light and darkness _ what kind of darkness? A vacuum of manifestation, of mental clarity, images that they override the information and at that moment a light beam is crossed by thought, an encyclopedia of information rises and the mind can spend hours retrieving information on a specific topic. With the focus on a specific that it will join a string of related information. But it is not the event itself, it is the chain of events that crowd and attract other creative powers comprising the rest of the elements that are associated, by any order that we call divine or harmonic phase of life of what would then be the human being manifests. The same evolutionary in what conditions are recreated then it would be the variety of the kingdoms that emerged and its derivatives, to overcome the pitfalls of physical reactions. The first thing that manifests itself as a primary law is protection of this source of energy that is beginning to act, with the things that surround it and with which it interacts. Comes an outer layer that protects it from damage making dense or lighter, depending on radiation that she graduates and controls for its interact with vibratory rates required for building their structures and functions, or vice versa. It is the reaction to the powers that penetrate it to promote its influence in the matter: his generation of the original Kingdom, an implant himself in the thalamus as the divine substance Dios in it is Earth. The insulation of the creative force that I discussed early on I am trapped inside that cockpit which ended the influence of external forces. Which were only necessary for its production of its components? "The camera bridal"-the Eva and Adam were prisoners of them influences outside. In principle paradise, then the degradation of the Kingdom which became embroiled in the pure being distorted itself and strangle, divided his body into male female. The loss of contact with the pure energy and unique condition that ancients discover the loss of alignment with the forces of the same system that would house them. The alpha and the omega: the creation starts the communication through networks that emerge from its Center, nerve communication terminals. They arise from the own need that creates them the environment and its components. That first egg of the reaction

78- Rene Quinton - Dr. Linus Pauling

The origin of being in salt water that before he presented conclusions on the same subject in the year 1904. Linus Pauling: Postulated that the human being has 118 elements of the periodic table in your body, which are present in water body's salty seas and lakes. Also, announced that cell originated in the Precambrian 3,800 million years ago, in the salt water. Another finding is that human plasma is like sea water, into its components. Note - with the variation that human beings now consume fresh water and should have a slight variation in composition, although it from consuming processed salt. The cells of proteins that are starting to produce; acquires some process of attraction or magnet inside what we call a conscience or individual intelligence, or need for attraction to other elements which contribute electrons and complement their development. This quality or notion of matter and its components, is both a divine intelligence that emanate towards creating and isolated herself in a cabin where started the control and the stages of creation of its own being. There are countless parts that come to organize this process through stages of time and addition, per emerging needs to create. An energy that envelops in magnetic and gravitational fields gives the primary creation comes the ability to move and begins to move in their environment and develop increasingly complex qualities. As the seed, whenever it acquires a new aggregate copy it somehow to their composition and keeps it as the string that is formed in its evolution. Or that, it creates a copy of energy and duplicates it in any other division that believes. The copies that will inherit those genetic codes to divide it, which is the law that formed them, pass them to their duplicates. This form of division an interact creates that, it defines a realm with unique features. By deduction every created thing duplicates itself. It emerges in another stage of manifestation with the above features or a clone of its predecessor. If that is the event of evolution, because it, who denies the reincarnation, needs move create them bulges outwards. At the same time, own gland provides sensory branches of them to detect nutrients and forms which must avoid and overcome in their movements. To these features is ads a skill to copy and retain a memory of them forms and energies of it detected for in a future catalog and identify the experience. That would be recognized if manifests itself again and the reaction of per recorded information and the notion of your choice or rejection of each to keep a code to play which were necessary to overcome a condition of growth.

At the same time, they must face on substances that are added to the body by food to suit their conditions to train the system to produce more cells with those subjects which are of different qualities. It is possible that in the long run is so much variety which must modify their behavior according the regions and traditions on which they feed. By becoming more complex processes that must be overcome and additions that come are very weak due to the concentration in water, as all created beings begin the evolution of their senses of movement, to create what would be the eyes to perceive light and movement. A skeleton that supports the ravages of the calamities, the time and the waves that responds. This would use the same scheme from the shells to protect sensitive organs. While skin protects organs and sensitive parts, the same time that acts as a filter for subjects in activity. Internal glands to support more creative energy needs giving them an immanent need to generate other waves to provide adequate vibrations to create components that enable them to develop new areas of development and mutation. It sucked the nutrients that were adding to their structures, in the same way seeping fluids to retain the components in water, for their benefit was created at the time that they added a new feature to the previous Fund. Themselves stretched in their nerve fibers and taking awareness of the dimensions that include signals associated to new experiences. They were creating new nerve areas and cellular structures of materials that sucked to store that information and hence arises the first brain. In the cells, themselves, they doubled that genetic information, passed it to the copies, which survived did the same. Already being a cell with two divisions and functions, each side control has its genetic code. Emerging internal gender, where the needs are developed transfer genes from one cell to another for to complete the cycle of division, with appropriate polarities. The primitive internal gonads arise inside to train each duplicate codes of evolution. This process arises the invasion of other forms of life such as bacteria or viruses and strange shapes to their development which should begin to develop defenses for your protection. Of this interact begins the creation internal of what would be the prototype of the human being and an intelligence or accumulation of details accurate of it created. A hive of nerve reactions develops a center of neurons, which responds to impulses that receives and sends energy responses that are related and recreate a condition correctly interpret the impulses that collects. By the need to accommodate a structure that was developing its size, it was doubling its structure in what would be the body and through their nerve networks sent the necessary orders of construction of attributes that would play its primary structure in a body. It gave him of communication networks, where he attained full control of each cell created in the interior. He sent instructions through neurons or nerve centers created to maintain full control of every action that you gave in your body. Maintained and maintains a notion, as an exact copy of your content. If any cell or group ceased its functions is destroyed, the matter to replace it was immediately generated and sent to the right place. He created a center or factory of cells in what we now call liver, which creates the exact component, to maintain the active building needs in harmony with other fluids related subjects. Somewhere in your brain, it developed a quality of generated pulses, orders making duplicates of already created cells. As always, in their primary cells duplicate or copied a replica of its advances in the dating of his new creation. Begin a palpable fact that that set of nerve cells, no matter the region that has been sent, forms a network of internal communications. The cells are aware of the functions of the other, forming a consciousness or universal mind than they are and their relationship to the whole. This intelligence is part of the control unit, the thalamus and endocrine glands that accompany it in its functions. It is a perfect operations complex. How might an impulse of pain take an accurate statement to this Center and give knowledge of any illness or injury, in which cell or cells there is injury or condition, that exactly the material required for your prompt service is dispatched from his laboratory to the precise site where it is needed in the precise amount, direct it through all its channels to the sites where the damage is and duplicate it exactly to the former? But there's more: If for some reason this material, either by the rupture of a vein or capillary which leads it, torrent of liquid that directs it to your destination in a part where is not recognized and does not reach its destination, the brain immediately sends new cells to create a layer to its around and insulates it creating what we call cysts. This central is seizes of the knowledge of it created and we are aware, we give has of that intelligence; alerts us and keeps us aware of its functions. That same area, the thalamus, is the part where has recorded them profiles of all what is perceived and creates a map, as a GPS of the worlds and experiences to which has had access, a Bank of memory of all to what has been exposed. It gives us the freedom of movement and to choose situations that are harmonious with their knowledge. If you detect any difficulty, alerts us to avoid it or generates impulses necessary to adjust and overcome it. This is an imaginary outline of how did one of the kingdoms, the most complex of the creation: human and divine light, which produced it. Intelligence that adds memories to all what is done and somewhere is recorded as an inner world that takes knowledge or awareness of his being, he realizes everything and you can say that. Concepts and stereotypes of human knowledge, is in the branch that manifested, may not include the complete plan of creation. Behind all this notion there are worlds of knowledge that are not or can be captured by the human mind. We can lock up in laboratories and see each individual cell and decompose one to a part, either atoms or all its particles, either in energy or in its derivations, in the way that each be wants to study the composition of the bodies and understand their aggregates, their

DNA chains. Of one thing, I am sure: that none could give a description and a concept that encompasses everything that involve being divine and internal processes. The Gnostic knowledge are the only internal source to harmonize our being rational with mystic and divine attributes which dwell by spiritual inheritance in our interior, the maturation of our soul and the conscience. The intuition of the God of our consciousness lifts us to a perception of States of exaltation to meet that divine harmony. The wisdom that was present in the first movements Essenes and the mindset of that period could not sustain by the poverty of preparation and imagination, where the deposit of the soul was attributed was behind the ear and stuff like that. The paradox of the largest, if so it can be called, is for me to prove after arise the first cell of the combination of sodium and potassium and other aggregates, then, by evolution, giving place to the thalamus. Accordingly, if the be proceeds of this first creation and depended on of them elements sodium and potassium concentrates in the water salt: how is that leaves of its first source and continues more afternoon around the land, and then depends on of the water sweet for follow subsisting? As you know, when creating the emanations of the Sun was possibly very concentrated and elements accumulated in lakes and seas. Water, filtered by lots of Earth, purified, while being developed and gradually was interacting and gradually migrating between the salt water and the Earth, and then serves freshwater subsistence. Not only the human being developed this application, others came to the same stage of evolution as human beings. The logical explanation and was not within the reach of the human mind was that the fumes of sodium and potassium necessary for survival came from the sun itself. So, that there is a correlation between sodium and potassium that are contained in the water and sodium and potassium that emanate from the Sun, as well as several elements which focused on different atmospheres and matter than was Earth. Any transformation emerges in the membrane to fit the two environments and adapt to direct oxygen from the environment out of the water, the membrane that exerts this work during the period that was immersed in water, gradually disappears when he is adapting to life outside the salt water. This gives rise to material life in this way possible both on Earth and in the sea. The thing that most strikes me is that in the beginning came a Flash of creation where was born the first glow of creative energy, a phenomenon that has not be repeated in the face of the Earth. Everything that emerged in that brief period simply manifested itself. All forms and functions that inherited the creation in the kingdoms that preceded it came from the demonstration. The created realms what they have done is simply evolve a first demonstration and vary their stages of evolution, surpassing the first stage of creation for thousands of years. Simply, this creative impulse took office, with attributes as each party that generated, the qualities that gave each creature, from the very beginning. The original purity of the creation has not recurred since emerged the creation on their debut... This deduction comes from the reasoning that, if the same condition emanates constantly, as in origin, the elements that came together to that exists, would be creating every day new forms of life. What we see is that the original forms evolve to more developed States of their own and original creations. Creatures and plants arise from its predecessors, inherited conditions that have already been part of a previous life. These new developments from the previous duplicate virtually characteristics of the preliminaries. But how come from nothing something with totally new features, demonstrating an unknown? It has not occurred in this earthly plane, which is known in the creation and we found the strands of DNA that follow a hereditary pattern. What generated those strings of inheritance? _ Would in substance or essence was generated the first creative impulse, the thing, as it is called, which interact in the first moment that life arose? _ What chain of emanations binds us together in a common knowledge of all phenomena that clothes all what is known? Would of where the gradation of emotions of them more different that we characterize as race human? _ Would in what scale is can catalog and measure these vibrations that not obey to them laws materials and that is projected in the scales sublime of what we perceive? Why not can the infinite variation of these scales be captured completely by the human mind? _ What is the harmony of creation and what plans belonging or is endowed, and why and for what? We know and are aware that affects us directly, and cannot, and imagine its real functions. Our imagination creates, and is part of them, but we do not master it. We draw energy from the cosmic plane, entering service with our energies, we call spiritual as physical, and they build within ourselves, we don't realize their specific functions. Simply, we know that they work, but we cannot label those functions to capacity. We use phrases, verbs, all the lines and cannot even evolve a clear idea of these functions. If we abandon them, they follow their duties unless we realize. An interact energy at all levels, which combine the products processed by our glandular system and the product is added to our physical body while it was part of the frame of the cell and energy structure of what is an individual personality. A joint so exceptional of energies, of all the classes that not mastered and are present in the life simple or complex, us fogs the understanding. Simple question: we are on the doorstep of the most extraordinary revelations that I just translate this written. Teaching methods that gave to the disciples and teachers of a superior in the realization of that knowledge were used in the ancient schools of learning in the countries of India, Egypt, Greece, and Mesopotamia. _ How the human being, through voluntary processes, could influence the processes of creation and training your body and mind too higher in these energy management functions? _ The laws of resurrection, travel astral, the spirit flow out of the body and return- the reincarnation, where perceived that the new be inheriting of his previous life qualities of the soul personality that not is destroy or lose with the change. Showed that in a new being manifested something of a previous and who simply followed its evolution to more advanced stages of knowledge and maturity. It became known as the esoteric hermetic schools or of the mysteries.

After millions of years of living and have developed a system quite advanced and complex where emerge vision systems by the emission of light and many qualities that were perfected in accordance with their survival needs, the fact of living in an area of water trained them in ways to breathe in water by ever shorter periods. Interact in the water to the land and vice versa created alternatives to survive. Being manifested in the placenta where the creatures are developed. This playback system has not changed never until today: being is maintained in a bag of fluid. A membrane created to protect a body to the same capacity that is created the protection of the first cell where is content a duplicate of the creation. These are universal laws of creation and the preservation of the minds that created them, gifted as those functions as delicate as his own emanation processing of an Earth-wormlike. At present, the earthworm has more than one heart - some, seven - and is a simple model of what our internal system of processing them. Other surrounding systems complement livelihood, such as the lungs, which increasingly acquire more capacity to filter the air. In the head is manifest a membrane in the creature newly born that breathes by it by a period of adaptation. Model of its existence immersed in water. They have done experiments where the creature can follow submerged without breathing through the nose or mouth after birth. I even remember my mother blowing this area children choking. This gives us a notion and proof that remain vestiges of changes that are operated to migrate from water to land. It is a quality that also manifests itself in another known species.

87 - The path of divine consciousness

Each nerve strand runs along the boundaries of our physical being, establishes one of the most perfect communication networks that human beings can imagine. They penetrate each terminal and every corner where some function is in process. They function as the roots of plants in nature. Its function is to deal with a specific area to establish another center of sensitivity, which collect information and return it to the command center, which is the brain, thalamus and its secondary glands. But this not ends there: are responsible of carry to all the components of the creation them data and instructions for activate the sensitivity that us gives a reaction of protection, in case of danger; In addition, if an injury or disturbance by disease, they activate other control centers so that they generate the reactions necessary to establish harmony and preserve laws operating at full capacity. This type of control, human beings had not survived the amount of time it has achieved with the evolution. Only our subconscious gives account of activities that we can access internally. A clear example of the power of these internal energies that may be manifested is related to epileptic seizures. Downloads are so powerful that make getting nerves and muscles of the body, where the contortions of the affected are manifested. The same reaction is observed in the flesh Gets an electric shock. The structure complete of the brain derives from endocrine glands that respond to the needs of survival, in addition to all functions complex that has developed, the human being, controlled by this small center of communications, in association with the thalamus and energies which penetrate our system, loading of atoms and reactions which we do not control, who travel to all corners of our bodies and out of our body. The first cell that begins to put together a cluster of cells, which form a mass or body, becomes then the thalamus. It develops a kind of foot or opening that springs from a central mass that used to dig and hold. This bump which has a hollow tube served him to navigate and detect movements of other species, at the same time, to suck matter that he combined with proteins to continue creating what would be a body. They developed brain mass, pédicas and visceral. Through hollow extrusions developed a nervous system consisting of neural systems and networks that interact with each other. He then created an extension which took place in what was then the body, an aggregate of mass to form a more complex structure that would supply their needs, where he created a structure to respond to their needs already in process of creation, being. The heart, in a transparent bag, has pericardium and Atria of irregular shape, and ventricles. Developed aortas, which carries blood to all the body; the blood returns to the heart via poisonous system. This little scheme of evolution, functions appears to the development of the human being to nature could express themselves, in principle. Our evolution must have been a different and more complex process, but the processes of evolution are consonants to examples like these.

89- The heroic age of spectroscopy

The discoveries of this French scientist that corroborate the theory

Experiment "the striped black of the Sun". When the darkness comes the light, or complemented (page 73: A bit of science for everyone) Newton had shown that light was the result of the combination of the seven fundamental colors. The refraction through prisms decomposes white light that enters our system in a spectrum reflecting the seven colors of the Rainbow. In the experiments of Bunsen and Kirchhoff to demonstrate refraction of sunlight and why produced fragments of light too dark in the solar spectrum, throw powder of sodium into a flame and analyze using a Prism the emitted color was verified that sodium is equipped with features, well-defined stripes. Doing the same experiment with potassium, they proved there are stripes, but are different. They studied all the bodies chemical possible to analyze the chemical constituents of the Sun and, later, of distant stars. So were born the Astrophysics and cosmic chemistry. In the experiment of Fraunhofer, identifies the black stripes of the solar spectrum, what was its origin. Fifty years later, Bunsen and Kirchhoff noted that, among the rays of the solar spectrum, two of them correspond to the stripes of the sodium that had been identified in the laboratory. Therefore, they decided to put in front of the Prism with which analyzed solar light a flame with sodium. To his surprise, the two black stripes were blacker. Kirchhoff moved then to the hypothesis of that the sodium of the calls had absorbed them stripes issued by the sodium solar. They discovered that the sodium emits and absorbs the same stripes. If there is passed to another deduction: if those same stripes of emission of sodium exist in the spectrum solar normal, between the Prisms there is sodium that absorbs those same stripes. They attributed the absorption of sodium into the Earth's atmosphere. It would be a mistake, because if the absorption of the rays comes from an atmosphere, that is of the same sun. Put another way, if the solar spectrum contains black stripes is because the chemical composition of the solar atmosphere filters, eliminates and absorb certain rays of light emitted by the Sun inside. The Sun emits light and, to its time, its atmosphere absorbs part of it. Therefore, solar emanations to the land and water include sodium and potassium is concentrated in the two bodies.

It is reproduced below, the chapter "The heroic age of spectroscopy", from the book history of the physical, Desiderius Papp (Espasa-Calpe). The best telescopes of the time came later, at the hands of this great master of optics. During a trial with a prism of exceptional clarity, Fraunhofer stumbled, in 1814, with the largest find of his life. The Prism had spread the solar light in a spectrum that the Sage observed through the telescope of a theodolite. He was surprised the spectrum vertically through many dark lines. By the way, it was not the first to see these enigmatic striped; two years before that he, the physicist and chemist English Hyde Wollaston (1766-1828) perceived them throughout the four main colors of the spectrum bands. Took them for the dividing lines, separate nuance of another, not paid attention to them Fraunhofer had more than five hundred stripes, and appointed to the most apparent with letters of the alphabet, thus creating the basis of a nomenclature which his successors did not have to expand. Each stripe, he acknowledged, has an exactly determined irrefrangibility. Examining the lines for different positions of your device and different positions of the Sun, he saw that those not moved; probably, he suspected Fraunhofer, are inherent to the same source of light. His interest was powerfully stimulated now: did pass the rays of the Moon and the planet Venus through a Prism. Their spectra are crossed by the same lines that had entered in the light solar. The spectra of several stars' parade on the screen in the following months; some of these stars, like the goat, are reproductions, though weaker solar lines; others noticed a different design. Just at that time, to 1818, arrive from France news about investigations by Fresnel; the obvious interpretation given by the French physicist to diffraction deeply impresses the Bavarian optical, firmly convinced of the wisdom of the wave theory. In a series of observations, replaces the Prism plates of glass and metal, which has drawn close stretch marks from the other, up to 300 in a millimeter. These diffraction networks serve to measure wavelength of the dark lines in the spectrum. Admirable is the accuracy of your measurements, whose errors are below 1 per 1000. Fraunhofer was not limited to watch the celestial light sources; the flames of the spark plug and oil lamps presented continuous and unlimited spectra. Not escaped your attention that the introduction of a salt in the flame did appear in the spectroscope bright stripes and saw the yellow stripe drawn by the flame of the sodium is split into two lines passing through a most powerful Prism. You searched for them in the solar spectrum and soon realize that double stripe is marked exactly in the same place where two black stripes were in the solar spectrum, that had intrigued him since the day of its discovery and that he had designated with the letter D. presage the importance of the enigmatic coincidence, without being able to interpret it. Absorbed each other practical research, and the mysterious lines served as reference signals in their searches for the index of refraction of different kinds of crystals. Its arcane not disturbed him more. As Fresnel, Fraunhofer died of tuberculosis and to the same age that the great French: to the thirty and nine years. Three decades had to pass after the death of the discoverer before the Bunsen and Kirchhoff arrived to decipher the enigma of the dark Fraunhofer lines and create the magnificent instrument of exploration which is spectral analysis. During that time, many researchers rubbing the discovery, that always slips from your hands. Talbot and J. Herschel recognized that the same substance that colors the flame of the spirit always emits the same stripes. "When in the spectrum of a flame - enunciates Talbot - appear certain and certain rays, these are safe features of metal contained in the flame". Wheatstone extends observations by studying the electric arc light, and finds different stripes, per the metals used in the electrodes. Miller makes through sunlight by vapors of iodine and bromine to examine the absorption lines. Glimpses of the knowledge are born without a certainty. Foucault, so skillful in other experiments, groping this time in the dark and not overshoot the species of stripe sodium D. Follow the Angstrom Swedish and English Swan, Stokes and Brewster. The latter recognizes that certain dark lines in the solar spectrum are engendered by the absorption of the rays in the Earth's atmosphere. All these wiser only make findings isolated and incoherent; finally, in 1859, the capital discovery of Bunsen and Kirchhoff: set out with clarity the law and achieved the first applications.

The chemist R. G. Bunsen (1811-1899), as inventive as a tireless experimenter, and the brilliant theorist of physics G. Kirchhoff, (1824-1887), both professors of the old and renowned University of Heidelberg were supplemented in the happiest way and their collaboration could not be more fertile. Flame colored by given substances caught the attention of Bunsen, who endeavored to obtain from them a secure way to identify chemical bodies. Obviously, there was a need first, having a pure flame. It's the alcohol, with the inevitable impurities introduced by the wick, not paid; lighting gas seemed most appropriate. Trials of Bunsen to mix air with gas lighting without explosion came in 1884 burner that bears his name, a flame source constant, pure, without light, auxiliary indispensable since then in the laboratories. Bunsen is not content to observe the colors caused by different substances in the flame of his cigarette at a glance: examines them, following the advice of Kirchhoff, through prisms. The results led him very soon to recognize that bright rays emitted by incandescent and metal vapors are independent of temperature, independent also of the elements with which metals are combined, and offer safe and constant features of the chemical Corps, although they present in minimal amounts just less than a ten-millionth gram of sodium to produce the double yellow stripe. That continues to display still the presence of this element when the analytical chemistry does not discover the slightest vestige of this. The study of rays emitted by several bodies, in flame, either in the electric arc furnace or in the electric spark, they convinced the safety Bunsen of his method, very soon brilliantly confirmed by the discovery of two new elements. Rubidium and cesium, found by Bunsen in 1860 and 1861, respectively, received their corresponding names in the spectral lines that allow finding them. Emission spectral analysis was founded. It required to be completed to become an instrument whose scope once more, as in the time of Newton - ranges from Earth to the outposts of the sky. Transcendental amplification of the efficacy of spectral analysis meant the solution of the enigma still open of the Fraunhofer lines and was the work of Kirchhoff. Produce in the lab, artificially, Fraunhofer lines in the spectrum was the first decisive success which gave the key to the problem. Kirchhoff and Bunsen executed the feat so once seems very simple. Kirchhoff went on an intense called engendering of a spectrum continuous: in the way of them rays, placed a lamp of alcohol with solution of salts of sodium, CA of the feature double stripe yellow. Instantly, the lines yellow and bright is developed in lines black D, identical to those of the spectrum solar. If taking lithium chloride rather than sodium salts, he saw red stripe feature lithium become dark. It recognized that you simply placing colored flames, sources of bright lines between a sufficiently intense light source and screen of a spectroscope to see flames absorb the rays of the same wavelength emitted, and introduce the spectrum, instead, black stripes. "He said - he wrote Kirchhoff in October 1859 to the Academy of Berlin - the dark lines in the solar spectrum that are not caused by the Earth's atmosphere originated by the presence in the hot solar atmosphere of those substances that have bright lines in the same place in the spectrum of a flame. We can admit that bright lines of the spectrum of a flame, which coincide with the D lines, are always due to the sodium of these content. D dark lines in the solar spectrum suggest, therefore, that is sodium in the atmosphere of the Sun." As fumes from solar sheath are colder than the astral, a given element of the solar atmosphere is unable to replace their own radiation rays which has absorbed. So were born the dark lines in the solar spectrum, lagoons that translate the absence in light rays of given elements and its presence in the Sun. The enigma of the Fraunhofer lines was, therefore, determined, and at the same time open the possibility of the chemical analysis of the Sun, possibility considered some decades earlier by the French philosopher Augusto Comte as a dream out of human reach. More here did not stop Kirchhoff; two months after its first communication to the Academy of Berlin, proceeded to generalization and rigorous testing of the law which had been found. It introduced a new notion of the body perfectly black, susceptible to fully absorb the rays of all wavelengths and does not reflect any. Such body, a full radiator, there was at that time more than in the imagination of Kirchhoff, and conducted technically later, in 1895, Wien and Lummer. A time defined the body black, Kirchhoff showed the validity of the equality where is the power emissive, the power absorbent of a body either, and E and to them powers of emission and absorption of the body black. As they absorb all the rays, A is equal to the unit, so that the powers of emission and absorption of a given body e/a ratio is a well determined constant. And Kirchhoff law States: for the radiation of the same wavelength, at the same temperature, the relationship between the power of emission and absorption power is always the same. The idea, converted by the discovery of Kirchhoff and Bunsen, that is given to the male penetrate the chemical nature of substances separated from us by impassable chasms of space, seemed not only to Augusto Comte, the Prophet debunked, but the same feat witnesses, incredible and utopian. Fun is to read the words of Kirchhoff in a letter written in 1859 to his brother Otón: "my attempt, the chemical analysis of the Sun, like many very daring. I'm not angry with a philosopher at the University by having counted, while we find, that a madman pretends to have discovered sodium in the Sun. I could not resist the temptation to reveal that this loco with the law of Kirchhoff, the interpretation of the Spectra received a solid foundation, and the decipherment of the spectral signals failed to start, supported on the one hand by ever deeper knowledge of the spectra of emission of chemical elements, and on the other hand by the growing power of the appliances. To the spectroscope of Kirchhoff and

Bunsen is associated with the network of diffraction; progress of splitting machine, the American physicist Enrique Rowland created in 1882 networks formed by ridges of surprising subtlety, up to 1100 in a millimeter. Rowland also applied the division into furrows to concave mirrors. Kirchhoff drew up a map of the solar spectrum, assigning many lines the chemical elements that engender them. A. J. Angström Swedish followed him; He was the first who described the solar rays in terms of wavelength. In the same year, 1868, the English Astronomer G. Huggins directed the spectroscope toward Syrian and applying the Doppler effect measured the bleed lines, caused by the removal of the star. Thus, evaluated the radial velocity of a star for the first time

Few months before, still in the same year of 1868, an eclipse total of Sun gave a clear test of the certainty of the discovery of Bunsen and Kirchhoff: during few seconds, it photosphere of the Sun was covered by the Moon and suddenly appeared, instead of them lines dark, the corresponding lines bright of the spectrum lightning issued by it atmosphere solar, thanks to the eclipse, was the only one who shone. A new science was born: The Astrophysics. In the decades that followed the feat of Kirchhoff and Bunsen, known as science put, increasingly, the scope of the physical-chemical exploration, not only the Sun and stars, but that also the spectroscopic eye penetrated up to the inside of the nebulae, away from the Earth by millions of light years. Contrary to it expected, no body chemical unknown in the nature land drew their stripes on them plates of them by the English J. Lockyer and attributed of first intention to an element that only exist in the Sun, is ended by find it (1895) as forming part member of the atmosphere of the balloon. Spectral analysis revealed the chemical analogy between the stars and rose to the rank of certain substantial concordance of the Earth with the remotest stars of the Milky Way and even distant galaxies. The display of the unit material of the cosmos searchable is the sublime lesson, historically the first that we were granted by the spectroscope, thanks to Kirchhoff and to Bunsen. However, this success, despite the magnificent, is only one of many aspects of the knowledge opened by the decipherment of the spectral lines. These supplied also posts by processes in the Atomic Mechanism Begetter of the spectral lines. They are like the echo of far configuration changes to satisfy in the universe of the infinitely small. Almost all the progress made during the Twentieth century, in the exploration of Atomic inside, then we must in-depth interpretation of the spectral lines. Have extended the scope of the investigation at the same time to the outposts of the macrocosm and the no less unfathomable depths of the microcosm is the significance of the work by Kirchhoff and Bunsen, comparable, in its majestic scope, to the discoveries of Newton! The beautiful simplicity of the spectra, as demonstrated in the experiences of the two initiators, due soon give way to the understanding of the spectrum depends on not only the bodies in the presence, but also the way in which they are excited. The spectrum of a given element changes as it is vaporized in an arc furnace or radiation excited by electrical discharges. Added to simple flame, Spectra from arc and spark, higher temperatures than they, the last studied since 1865 by Julius Plücker and Guillermo Hittorf Germans. Here he began a long series of descriptive work to accurately fix the emission spectra of the different elements, some of which, such as the iron, revealed its extreme complexity. Impossible is here follow the Chronicle of patient and laborious investigations which led, especially by Keyser and Runge, in Bonn, and later to Exner and Eder, Vienna, real encyclopedias of the spectral lines. Once measures, after a giant, spectral ray, and assigned to each item, arose the question of if the distribution of the properties of a given element lines, scattered along the entire length of the spectrum, is not subject to a rhythmic order. It could boast that a certain periodicity was itself. A vibrating string stored in its sounds a certain number of notes that can be covered in a formula. The simple formula, the theory had established for the vibration sound, would be impossible find it for them vibrations bright? The Swiss scholar J. J. Balmer (1825-1898) was not the first to propose this task, but its imitators did not have endless patience or shared its unwavering conviction that sought law existed. Balmer, master of drawing, as artist as Sage, was persuaded of the omnipresence of harmonic relations in physical phenomena; not admitting that the spectrum might be an exception. Swiss perseverance eventually succeeds in 1885, when he stumbled, after many calculations, numerical relationship that governed between stripes of hydrogen in the visible part of the spectrum. Balmer empirical formula describes with extraordinary accuracy the wave length of the stripes of the hydrogen, where a constant and m is can take integer values from three. Kayser and Runge replaced the law of Balmer wavelength frequency and obtained the formula that translates into the current notation by: where R is a constant and an integer greater than 2; each value is a line. The frequencies of the hydrogen lines are admirably due to the Balmer formula. In its discovery knowledge that the Swiss researcher was far from suspecting they were hiding, also. Their discovery soon became a true instrument of prophecies. The formula, already widespread in our century by Walter Ritz (1908) allowed to foresee not just one, but a set of series of stripes of hydrogen in the ultraviolet and infrared spectrum. The experience has justified magnificently forecasts and at least in the spectrum of the simplest of atoms, hydrogen, chaos gave way to the rhythmic order and all their stripes gathered in a formula were subjected to the law of 9-Balmer-Ritz. In addition, it was revealed that lines of other elements are also alike, although most formulas complex. They are representable by differences of quadratic expressions. Particularly, the constant R of the law is in the series of spectral rays of all elements; It is a universal and fundamental data as it has shown the Swedish physicist Rydberg, whose name was linked with the constant R. To establish a fixed relationship between emission and absorption of radiation, Kirchhoff opened, as we have just seen, the road to the magnificent emergence of spectroscopy; his law clarified many problems, but he also births others. Black body's completely absorbing rays of all wavelengths emitted them also of all being, therefore equipped with maximum power emission. This depends only on the temperature. What is the law of this unit? The law links the total radiation of blackbody temperature. Supported measurements J. Tyndall and others, the Austrian physicist J. Stefan (1835-1893) deduced in 1879 that the total black body radiation is proportional to the fourth power of its absolute temperature. Once determined the number of calories radiated in one second by a square centimeter of blackbody, Stefan law allowed to calculate the temperature of the Sun, in round figures, in 6000 degrees Celsius, on the condition that the Sun is a black body that absorbs all radiation, condition that seems, per recent experiences, in accordance with the reality. How is radiation from the black body on the different wavelengths of the spectrum distributed?

This problem already worried Kirchhoff. If is heated a piece of coal or iron, next to them rays infrared and red, that are the first in appear, arise with the temperature growing yellow, blue and violet. The domain of the rays emitted is moved, then, the low and high frequency. G. Wien studied this relationship, finding the law of bleed that bears his name in 1894: with increased temperature, maximum radiation intensity moves from the larger to the smaller wavelengths, so the product of the absolute temperature by the wavelength corresponding to the maximum is a constant. Stefan law is an empirical finding; the Austrian Physicist L. Boltzmann gave the necessary support, in accordance with the electromagnetic theory of light, solidly built by Maxwell. More the success is revealed well soon precarious. None of the thinkers of the 19TH century that turned to the problem of black body radiation could give a satisfactory interpretation of the spectral distribution of energy. While the curve feature in the campaign obtained experimentally introduced a maximum whose position was regulated by the Act of Wien, theory required a curve whose coordinates grow up to the sky, when wavelength increases. Nature reveals once more that their laws are not always accommodated to the reasoning of our spirit. Only the 20TH century released to the physics of the impasse that drove it the obvious contradiction between theory and experience. With the new century is born the new doctrine; on December 14, 1900 suggests Max Planck (1858-1947) the innovative idea to consider radiant emission as a batch process that occurs through elements isolated from power, holders of a certain magnitude. Such element, the quantum, it is proportional to the frequency of the beam, the proportionality factor being a universal constant of nature, the famous constant h who would later immortalize the name of its discoverer. Thus, the energy of an is given by the formula. The lucidity of this thought clarified slammed the enigma of radiation from a black body, immediately explaining the variation of the curve in campaign, whose whims had baffled investigators. Such success was no more than the first exploit of the new theory. Planck hypothesized was hiding the fertile seed of most of the outrageous and wonderful ideas that should transform, until it unrecognizable, the image of the physical world. One of his most resonant wins should be the explanation of the spectral lines by changing the interatomic electron configuration. Balmer Ritz Act admirably described the stripes of the hydrogen, but reveals nothing of why an element radiates a certain line and not another; He left the mysterious tie that binds the radiation with the atom radiator completely in shadows. Only when the Danish acute Niels Bohr (1913) introduced the quantum Atomic, giving to the circulating electrons governed by Planck's constant paths inside and when it was, with a bold hypothesis, electron to emit light to jump from one orbit to another, managed to get the frequency of the emitted radiation energy orbits differences. As for charm appeared in its calculation the frequencies of spectral lines. But now the heroic age of spectroscopy had long belonged to the past.

The first gland of the creation processes sodium and potassium to create proteins as I explained in the previous paragraphs. Allègre gave the nail on the head with his observations of Bunsen and Kirchhoff. The only mystery I know I could recognize and attach it to the theory proposed in my first book, by the pen of the bird and dry flower. The Sun emits energy that radiate toward the ground particles of sodium and potassium, the basic elements of creation, along with phosphorus, Sulphur, carbon, oxygen, hydrogen, salts and oils of the composition of other elements that come into play in the evolution. Are elements that are concentrated in the waters of the lakes and seas, and contained in own land? The first stage of evolution of the self is based on interact these elements and an energy that bounced off the elements sodium and potassium, and activated or ignited the spark of life on planet Earth. It is possible to operate the same throughout the universe but nature creates and manufactured with materials and elements by aggregation supplement needs. It is also this aggregation of materials that create the inherited qualities and then manifested in the created thing. Is a quality of the creation the Add materials to the already existing and generate new forms and energies? The oogenesis is currently studying is a phenomenon that has a pattern like this stage of early development. Possibly, as happens in other species, human beings had the ability to change sex in their original stages. It is a quality of some species that exist from millions of years ago, Simple change of sex or reversion, deletion from one to another, was a quality which became human in the stages of incarnation, where the quality sleeping during a phase of life arises on the other and vice versa when it is regenerated to be in a new body. The proof is that in clitoral area is that female hormones, created male or female member by selecting.

My personal analysis based on findings and information provided in these documents is an observation about recent discoveries in California, where there is a cell created and which evolves with arsenic. Normal would be that it developed with potassium or sodium, as in the first cell that emerged with the qualities to survive with the materials that we have human beings, of origin that we fear. The observations I have made in these findings suggests that there are potassium arsenate and arsenate of sodium, since he has denounced in the coasts of Mexico and other countries, the presence of these compounds. If by any chance an elementary cell of this compound has been developed, it would be another example of evolution, such as in the human thalamus. But I am concerned that the remarks only point to a single element in this interaction and it does not comply with the law of division of creation, because it should be compounds which give origin to vapors of life. I understand a simple law of creation and am that two conditions must be present to make emerge a third. In metaphysics, it is the law of triangle of manifestation, therefore, microbes, plants and animals; they can convert all these chemical compounds inorganic arsenic in organic compounds, committing atoms of carbon and hydrogen. The spark of energy, one of them, the sodium and potassium that I act with first energy radiation. It is a new element that provides a recent test by being discovered about a year ago, and its effects have not been revealed to the present, they are evidence that salt water you can recreate intelligent life. I may be an isolated phenomenon which occurs now, but God knows what time this is happening. The interesting thing would be to check if some form of intelligent life has emerged on that coast, which relates to this phenomenon and to verify in the future the Flash forms that shine in the knowledge on the subject. Those compounds: techniques of analysis chemical of the training of laboratory chemical of the College of Bachelors of the State of San Luis Potosí. Site 28.

Potassium Arsenate- sodium arsenate.

Introduction:

In several countries of Latin America, such as Argentina, Chile, Mexico and El Salvador, at least four million people drink on a permanent basis water with arsenic levels that put their health at risk. Concentrations of arsenic in water, especially groundwater, have levels that in some cases up to 1 mg/L. In other regions of the world, such as the India, China and Taiwan, the problem is even greater. The information obtained, in the India there are about 6 million people at risk, of which more than 2 million are children. In the United States, more than 350,000 people drink water whose content of arsenic is greater than 0.5 mg / L, and more than 2.5 million of people are being supplied with water with tenors of arsenic higher to 0,025 mg / L. The problem of arsenic in drinking water is trying to in the Argentina several years ago, when epidemiological Córdoba and other provinces of the country they showed and associated disease of doing, damage to the skin, with the presence of arsenic in drinking water. Efforts and studies performed to minimize it or eliminate it have achieved a breakthrough at the level of the water treatment to urban scale in the Argentina, Chile and Peru, but at the rural level, the solution in these countries is still pending. Hence the authorities of health Argentine promote with decision studies that involve a proposal for the solution or minimization of the problem designated. Effects of arsenic in humans is known as the main routes of exposure of people to arsenic ingestion and inhalation, which is cumulative in the body by chronic exposure and to certain concentrations causes conditions such as: disorders of the skin, relaxation of the skin capillaries and expansion of these, with secondary effects on the nervous system; irritation of the respiratory, gastrointestinal tract and hematopoietic organs; and accumulation in bones, muscles and skin, to a lesser extent in the liver and kidneys. There is epidemiological evidence that people with prolonged ingestion of inorganic arsenic, via drinking water, develop Hyperkeratosis span plant, whose main manifestation is the pigmentation of the skin and calluses on the palms of the hands and feet. Arsenic in natural waters; arsenic in surface and ground waters. Arsenic occurs naturally in sedimentary rocks and volcanic rocks, and geothermal waters. It arsenic is found in the nature with greater frequency as sulfide of arsenic and arsenopyrite, that is found as impurities in them deposits miners, or as arsenate and arsenate in them water's surface and underground. Arsenic is used commercially and industrially as an agent in the manufacture of transistors, lasers and semiconductors, as well as in the manufacture of glass, pigments, textiles, paper, metal adhesives, preservatives of food and wood, ammunition, tanning processes, pesticides and pharmaceuticals. The arsenic is present in the water by the dissolution natural of minerals of deposits geological, the download of them waste industrial and the sedimentation atmospheric. In surface waters with high content of oxygen, the most common species is the arsenic with ($As+5$) + 5 oxidation state. Under reducing conditions usually in sediments of lakes or groundwater, predominates the arsenic with ($As+3$) + 3 oxidation state, but there may also be the $As+5$. However, the conversion of $As+3$ to $As+5$ or vice versa is slow. Reduced compounds of $As+3$ can be found in oxidized media and $As+5$ media reduced oxidized compounds. Microbes, plants and animals can develop all these chemical compounds inorganic arsenic in organic compounds - committing atoms of carbon and hydrogen-.

105-Six- Creationist Conclusions

Conclusion: NO 1

Therefore, under the new discoveries of science and its exponents, it is possible to say that a new link in human knowledge can be taken as true. "The creation started and evolved into salt water"

We imagine that in the far universe are the clues to our origins, when I discovery that the energies that travel to our planet, are the courses of the beginning of life as local phenomenon, determined by the cosmic emanations to our environmental.

The conditions to creative life conditions are possible until now in planet Earth.

1. the writings of Dr. Esther Rosón Gomez the genetics of the brain and its functions, interact sodium and potassium as the main sources that gave him the ability to first cell produce proteins and create the first membrane of creation to begin the evolution and other features that tells us. The one who only exist as elements and be present for millions of years of existence, without any purpose defined as most the accumulated elements, in outer space and on Earth. That moment arises a new physical action of these elements, could be affected and penetrated by a foreign phenomenon which was not previously present, which activated the energy that kicked off this new manifestation. In some unknown region, emanates a subtle energy composition that altered the movement of the electrons of the primary matter to begin creating.

2 Claude Allègre, in a bit of science for everyone, US warns of the discoveries of Bunsen and Kirchhoff on chemical Astrophysics and its advances of the refraction of sunlight, which confirm that the sodium and potassium are the product of solar radiation and that our planet is penetrated by these and other elements that emanate from all cosmos that saturated the land and our seas from the primary era. And, by addition, all those scientists promote theories of our

system of energy internal, from Newton to Albert Einstein.

3 studies of the College of the Institute of chemistry in San Luis Potosí de México arsenic compounds, especially elements derived from potassium arsenate and arsenate of sodium, which combines with the findings of Dr. Esther Roson Gomez, which can be a source for future scientific research.

4. The new discovery of NASA on a cell that uses arsenic instead of phosphorus to reproduce. Perhaps they only gave with a link; there must be many out of human knowledge that has not yet been catalogued are like find an orchid in the swamp?

NASA Discovery

A strange bacterium can survive without one of the fundamental building blocks of biology. A bacterium found in the waters filled with arsenic in a lake in California is expected to give a twist to the scientific understanding of the biochemistry of living organisms. Germs seem to be able to replace the phosphorus by arsenic in some of its basic cellular processes, which suggests the possibility of a very different biochemistry we know that up to now, which could be used by organisms in past and present extreme environments of the Earth, or even other planets. Scientists have been considered for a long time that all living things need match to work, together with other elements such as hydrogen, oxygen, carbon, nitrogen and Sulphur. Ion phosphate, PO_4^{3-}, plays several roles in the cells: it preserves the structure of DNA and RNA, combines with lipids to create cell membranes and transports energy inside the cell through the molecule adenosine triphosphate (ATP). But Felisha Wolfe-Simon, geo-microbiologist and research fellow of astrobiology to NASA, the USA-based Geological Survey, in Menlo Park, California, and colleagues reported online today in the journal "Science" that a member of the family of proto bacteria can use arsenic instead of phosphorus. The finding implies that "potentially you can omit the match from the list of elements required for life", says

David Valentine, geo-microbiologist at the University of California, Santa Barbara. Many science fiction writers have proposed forms of life alternative, often building blocks using silicon rather than carbon, but this is the first case of a real organism. Arsenic sits right below phosphorus on the periodic table, and the two elements can play a similar role in chemical reactions. For example, the ion arsenate, As 43-, has the same structure tetrahedral and places of link that the phosphate. It is so similar that it can enter cells replacing the mechanism of transport of phosphate.

5-Tem studies of the DRA. Kristein M. Neiling on the influences of them energies electromagnetic of them planets on the behavior human, and its conclusion that this was something of which them old had control. Its beneficial effects of these scientific data and apply them personally. Electromagnetic fields of the Earth change for different periods of time. Today change began in 2005 and still it still affects electromagnetic Earth orientation. She showed that Earth is naturally aligned to changes. The same is true with animals. The only one that has that natural ability is the human being. The person being administrator of the astral and physical body functions should be on alert as understand its operation and upgrade it for the benefit of present generations. It belongs to health agencies function to orient and awareness of how to handle these data that directly affect the individual citizen, of the ways to harmonize their health with the right data. The attitudes of the masses in different countries are doomed to suffer from possibly catastrophic influences by these imbalances in physical laws. It is referenced to the disorientation and the current violence to these variations. This was a Commission from NASA to study the "Dead Sea scrolls" and its discovery, surprised, with the unusual discovery, a practice used by the inhabitants of those times use an exercise in concentration to balance the natural forces of being with these deviations from the matter. If you notice the position of the Egyptian statues with the usual position of the feet and the hands on the thighs that was a mystic stance carrying intended to maintain a balance of internal forces with universal energy. These new findings paid to the Declaration rudimentary, in that book and the theory presented in it. The purpose of exposing more elements that meet appropriate information to support this theory suggests to me that much relevant information already exists and that it has never linked to this phenomenon of primary evolution. If take it gland thalamus as the source of creation divine of them humans, must direct our attention, because that there is that is the energy required for orient it to them fields electromagnetic and achieve a balance glandular and hormone in our body.

6-The origin in the water salt Dr. Linus Pauling along with René Quinton, that before they presented some conclusions on the same subject in the year 1904. Linus Pauling: Postulated that the human being has 118 elements of the periodic table in your body, which are present in water body's salty seas and lakes. *Also, announced that life cell originated in the Precambrian 3,800 million years ago, in the salt water.* Another finding is that human plasma is like sea water, into its components. Note - with the variation that human beings now consume fresh water and should have a slight variation in composition, although it from consuming processed salt.

Conclusion: No - 2 are 6 tests and additional data that exert a powerful judgment on the exposed declarations where to be taken as certain in its details, is part of that separate the same conclusions, test "human beings originated in salt water". Gaspar (Edwin) Pagan

The human brain

A hive of energy from waves of frequencies and vibration on the scales of minimum and maximum rates being can imagine. The human aura is the emanation of the brain's energy and they transcend their cockpit to the outside, where they produce a halo of light that covers the seven colors or refractory of the cosmic spectrum white light. In elementary school, he would in the morning a small kiosk to buy sweets. I watched daily a huge pine that was left to the side and noticed that, in the upper part, which seemed the tip of an umbrella, it gave off beams of light color silver. Was as a flow of energy that followed its Dome where would like to that the wind it moved. A strange sensation of vertigo whistling through my body to look at the heights as happens to many people when they are in a tall building and facing down or up. I was amazed with this phenomenon. Never told it to anyone, because I didn't know what was. After so many years, I understand that he was facing one of the mysteries of nature: creation. Scholars mystics and scientists are wound his brains out trying to give the explanation of everything what is recorded and which refers to the mystical knowledge hidden in symbols that others understood subject matter have accumulated over the years around the world, hiding hidden meanings and veiled secrets. This was the Act of combination of energies of the forces of the cosmos, the Earth, the tree, which at the same time retained some of this for their growth and development, to serve as an intermediary and releasing subtle forces into space from Earth. Forces of electric discharge rays not have the subtlety of this pine. I also learned of this interaction that should have created a need to attract these energies so that life occurs. This is a conscious or unconscious of being need. Which manages to decipher his teaching and apply it will be in harmony with the divine creation: The Holy of Holies. Possibly, the same energies that flow through the dome of the pyramids of Egypt and penetrating the Interior of these cameras are and should have some sort of manifestation of this energy that was the secret knowledge that motivated the Egyptians to build so many monuments. Provable facts in the teachings of Dr. Linus Pauling, which relate the conclusions in my statements, are related to the energies that the Egyptians and other breeds used to harmonize your interior with the cosmic forces. The thing created in any of the kingdoms that is manifest establishes a harmony with the mind divine. It is the universal mind containing all the logos of creation, it, because it emanates from an active universal mind where all content is.

Do that consciously discover the channel of communication with the creative source will be an instrument of creation, a channel from that source to attract and emanate, as pine, the energy that compensates for the Act of creation. Upload and be part of it by time intervals and regenerated back to this plane of manifestation. After ponder all these characteristics laid down in this article and if any scientific logic is contained therein, may argue that any external or internal phenomena that changes the laws followed this evolution a phenomenon that could wreck the lives is complete or partial. Having said that, under the reasoning that any phenomenon that you alter the harmonious confluence with what already has been developed on this physical plane obeying these laws is logical to think that these same subtle laws of creation can be altered by phenomena external to the common manifestations of physical laws to which we are exposed. The internal systems of the realms of creation are overcome through evolution or overcoming internal conditions and external influences, which the body manufactures and adds to their structures is by any need of processing the advancement or change vital for their inner harmony. Any variation that is a violent changes their structures of DNA breaks the laws of evolution, therefore, his own life. The characteristics of any created thing succumb to change illogical. For example, just change a plant to a different climate makes that wilt and die, because it is has led him to a place energy that acquires non-harmonic with its development. Otherwise if you are in a poor environment and is changed to one more harmonious with its way of development this acquires a vitality that would give greater growth and vitality. Not only phenomena physical as the fire, the violence of those phenomena physical, can only destroy partially a thing, but then if is a vestige of seed this life will arise again. With only addressed a thought as simple as this opens a path for specular reasons that may affect the evolution. It may be a genetic phenomenon that occurs after a natural phenomenon.

The disappearance of the dinosaurs and other species can argue that they might suffer genetic changes in their DNA or your brain condition be affected by blockade of energy for your body evolution, to see their bodies slowly reduced by blocking substances, necessary so that his life would continue. The possibility hereditary that nullify the ability to reproduce or fertilize. The blockade of cosmic emanations of energy needed for your brain supplement the amount of energy, for the composition of their huge bodies and they deteriorate and disappear by genetic elimination. The dark corners of genetics and evolution have many loopholes; because it creates so much as destroy; only God knows the reasons. The fact that so many phenomena are manifested in changes in nature, the emergence of new species and the disappearance of others must lead us to reason in this way. The banks of corals and other species fragile merman slowly and studied its causes physical.

113- Second Part Breaking Myths

Pangea

Detachment of the continents

Uruguayan archaeologists have found fossils of almost 130 million years in the fossil deposits of the South American country.

This dating would give them greater antiquity of the remains of dinosaurs that inhabited the planet makes million years. This is say that they belong to the Gondwana, the southern continental block that arose after the first separation of the Pangea.

The second important stage in the break-up of Pangea started in the early Cretaceous (150-140 Ma), when the minor supercontinent of Gondwana separated in several continents (Africa, South America, India, Antarctica and Australia). Around 200 Ma, the continent of Cimmeria, as is has mentioned previously (see is "the formation of Pangea"), collided with Eurasia. However, a subduction zone was forming, as soon as Cimmeria collided.

Pangea

Archaeological test excavations showing data from the history of dinosaurs in the continent of South America have surfaced. Pangea is the phenomenon closer to the disappearance of these large animals.

Supercontinent that existed at the end of the Paleozoic era and the beginning of the Mesozoic, Grouped most the land on the planet. Was formed by the movement of tectonic plates, making a 300 million-year-old joined former continents in one single; later, about 200 million years ago, began to fracture and break up the current situation of the continent, in a process that continues. This name was apparently used for the first time by the German Alfred Wegener, lead author of the theory of continental drift in 1912.

The first continent that broke in segregated worlds, a debacle at the level of our planet after millions of years of existence, I think consequences that still the human being have not been able to decipher completely.

It comes from the Greek prefix "pan" meaning "all" and the word in Greek "Gea"
"Soil or land"

Uruguay South America

They are older than the dinosaur's fossils in Uruguay; it means that it is an event prior to the age of the dinosaurs.

Uruguayan archaeologists have found fossils of almost 130 million years in the fossil deposits of the South American country. This dating would give them greater antiquity of the remains of dinosaurs that inhabited the planet makes million years. This is say that they belong to the Gondwana, the southern continental block that arose after the first separation of the Pangea.

The latest find is a species consisting of two mandibular branches articulated; in which each has the trigeminal nerve and blood vessels. The paleontologist, Piñeiro explains: "We have not found a single reference that has previously registered a similar peculiarity and spectacular preservation".

This feature is what makes the discovery so particular. No fossil record has remains similar. For the paleontologist, the new finding is to go beyond. With new discoveries, experts may "go a step further" and learn about animals that lived makes nearly 300 million years in a hostile environment: "what we discovered allows us to know aspects of behavior that are little fosilizables. I.e., how they ate, how they reproduce, how are so well adapted to an environment that is not very favorable to life, as it is a little oxygen and very salty lake".

He places in which you have found them remains fossil is known as "Konservat Lagerstätte". They are deposits of fossils where, under very specific conditions, are conserved structures that normally would not do it in other places. The fossils found by the research team were in a set of rocks of more than 280 million years ago, called "I- Mangrullo". This would cover the areas of Tacuarembó and Cerro Largo, Rivera, although rocks extend to Brazil. Piñeiro commented: "we don't do excavation, work on Mangrullo rocks on the surface. These rocks continue in Brazil and for this reason we work together with geologists and paleontologists of that country" to move northward. In the Cretaceous, Atlantic, America and South Africa today, finally separated from Eastern Gondwana (Antarctica, India and Australia), making the opening of a "South Indian Ocean". In the Middle Cretaceous, Gondwana fragmented to open the South Atlantic Ocean; South America began to move toward the West, away from Africa. The South Atlantic did not develop uniformly; rather, rose from South to North.

In addition, at the same time, Madagascar and the India began to separate from Antarctica and moved northward, opening the Indian Ocean. Madagascar and the India separated each other 100-90 MA in the late Cretaceous. India continued moving is toward the North, towards Eurasia to 15 centimeters (6 inches) by year (a record of tectonics plates), closing the sea of Thetis, while Madagascar is stopped and is returned blocked to the African plate. New Zealand, New Caledonia and the rest of Zealand began to separate from Australia, moving eastward towards the Pacific and opening the Coral Sea and the Tasman Sea.

117- ARDI, the oldest human species

After the recent discovery, the site of Uruguay is the oldest in South America and the second largest in the world. The reconstruction of the place in which the animals lived 280 million years ago, is funded by the National Agency for research and innovation (ANII) of Uruguay and the results will be published in several scientific journals around the world.

It is described, gives us a clear notion of the phenomena that the planet Earth has suffered.

Discovery of ARDI a female of 4.4 million of years what today is Ethiopia demolishes the Darwin theory.

Our minds and history write it to those who have the power, if they seek the date in which emerges the figure of Darwin and his theory, we must bind created civilization and the domain of interest. Search in the basement of the creations of the less advanced is a maze for the common being that does not have the advantage of having an extraordinary fortune to create stories, before the cinema or television. The fantasies that are produced today take the same course; create truths disguised as to be common that cannot penetrate the truth. A child begins his life the greater part through drawings animated that it trains to believe anything that make a piece of cardboard, or drawing. Films have been made famous in that sense; everything is possible to create in the mind of humanity, because it has trained us to believe anything. No one questions the application of superfluous details while your mind is occupied. That has humanity with this practice. Every day being is farther from the truth, and the powerful can manipulate human intelligence.

This finding puts end to the theory of Darwin and to the idea of that the man descends from the Apes. In fact, the remains of Ardi offer a window into what the last common ancestor of humans and apes living today could have been.

Facts that abound in information that opens a new route to the science of the future to clarify taboos are given in different territories.

In the present continue them finds in the area and are have discovered many fossils more, both contemporary to Ardi and Lucy as also more ancient, as is the case of *"Chad" whose remains date back of approximately 6,000,000 of years.*

An international team of scientists presented which they say is the oldest fossil and best preserved of a direct ancestor of the human species.

ARDI, ARA-VP-6/500 catalog name, is the nickname given to the skeleton of a female belonging to the species Ardipithecus ramidus, probably a hominid (bipedal primate), which is considered the most primitive hominid known so far and that lived during the Pliocene, about 4.4 million years ago, Human species

It's a female of the species Ardipithecus an international team of scientists presented which they say is the fossil oldest and best preserved of a direct ancestor of ramidus, which lived 4.4 million years ago, in what is now the Ethiopia. As point researchers in the journal Science, although it was not a matter of our direct ancestor, the discovery offers valuable information about a crucial stage in human evolution: the moment in which we parted from the common branch that we share with monkeys. The discovery, say the researchers, shows like, never the biology of this first stage of human evolution.

Until now, the oldest known stage of human evolution was Australopithecus, the Biped's small brain. It is older than the famous of the australopithecines of Lucy, a 3.2 million years old fossil discovered in 1974 about 70 kilometers of where Ardi was found.

When Lucy was found the community international thought that the hominid more ancient would have an anatomy, like it of the chimpanzees, but Ardi, that is almost a million of years more ancient that Lucy, not supports that theory. Is it well similar? She was found in Ethiopia; it is a female of the species of hominids.

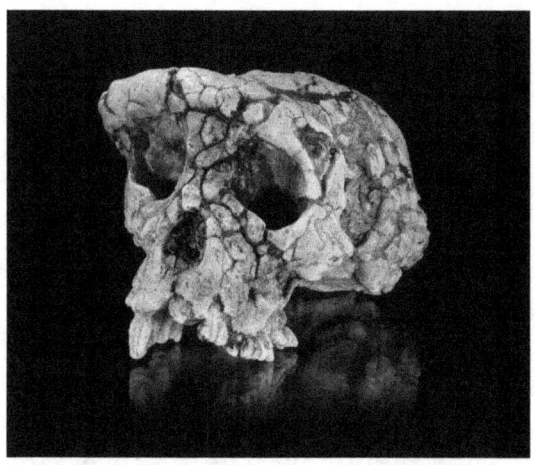

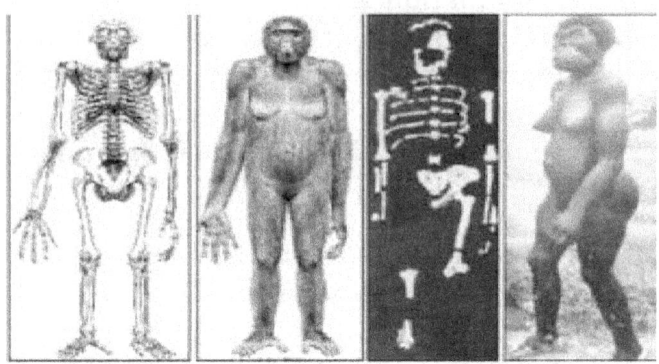

Fossil (Ardipithecus ramidus) tops the list of the most important scientific advances and is 1 million years old the latest found on the origin of man.

'Ardi' has more than one million years old that Lucy (Australopithecus afarensis), a partial skeleton of what was considered until now the oldest hominid.

The research of Ardipithecus has changed "the form in that think about the evolution human and represents the culmination of an arduous collaboration investigative of 47 scientific of nine countries that analyzed 150,000 specimens of animals and plants fossilized", said Bruce Alberts, head of drafting of Science in a publishing.

After analyzing the skull, teeth, pelvis, hands, feet and other bones of the fossil, the scientists determined that "Ardi" had a mixture of features "primitive" he shared with apes from the Miocene who preceded him and other "derived" characteristics it shares exclusively with hominids. The discovery of fossils done in 1994 revealed the biology of the early stages of human evolution better than any other to date, said the American geologist Giday Wolde Gabriel, who led the analysis of lavas and ashes that were used to determine the age of his remains. The whole of the fossilized bones of "Ardi" revealed that it was a woman who weighed about 50 kilograms and had a height of 1.20 meters. "With a skeleton, so complete and so many individuals of the same species in the same time horizon, we understand the biology of this hominid", said Gen Suwa, a Paleoanthropologist at the University of Tokyo and author of a report published by Science in October this year. ARDI, as it has been dubbed, was discovered in 1992 in the region of Afar, Ethiopia, but it took 17 years to carry out analyses of the finding.

The fossil has a small brain of 300 cm3, belonged to a female and was nicknamed "Ardi". Radiometric dating of volcanic lava layers revealed that Ardi lived 4.4 million years.

ARDI

Species: Ardipithecus ramidus

Sex: female.

Weight: 50 kg.

Height: 120 cm.

Preserved remains:

Hands, feet, legs, ankles, pelvis, and most of the skull.

The fossil reveals that the ancestor of the human lineage underwent a stage of evolution poorly known a million years before Lucy (Australopithecus afarensis), the iconic fossil female who lived 3.2 million years ago, and which was discovered in 1974, only 74 kilometers from the site where Ardi was found. The researchers argue that the shape of the pelvis, members suggest that it was bipedal when walking on the floor, but it was quadrupedal when he moved between the branches of the trees.

ARDI was partially specialized in the posture upright, with the waist pelvic somewhat "acuencada" to support them intestines in a posture vertical and with them bones of them feet slightly rigid to facilitate the displacement biped.

While she had long arms and curved fingers to hold on to the branches of the trees, it also had foot big toe or hallux, divergent from the other fingers, such as great apes, allowing you to hold with the foot, also to the trees.

18 OF MARCH 2015-14:05 APRIL HOLLOWAY

Other found cases

4 - Bulgaria

Sofia News Agency, Novinite informs that the skeleton was found in what was a giant human skeleton unearthed in Varna,

Archaeologists in Bulgaria have discovered the remains of what they described as a "huge" in the Centre of Varna, a city on the shores of the Black Sea, whose rich culture and civilizations extends for about 7,000 years. The ancient city of Odessus, since commercial establish by Greeks towards the end of the century seventh before Christ. Odessos was a mixed community consisting of ionic Greek and Thracian tribes (getas, Krobyzoi, Terizi). It was later controlled by the THRACIANS, Macedonians, and then the Romans. The Roman city, Odessus, covered 47 hectares in the current central Varna and had prominent public baths, Thermae, erected at the end of the second century of our era, now the largest remainder Romano in Bulgaria.

Read more: http://www.ancient-origins.net/news-history-archaeology/giant http://www.ancient-origins.net/news-history-archaeology/giant-human-skeleton-unearthed-varna-bulgaria-002787 human-skeleton-unearthed-varna-bulgaria http://www.ancient-origins.net/news-history-archaeology/giant-human-skeleton-unearthed-varna-bulgaria-002787 002787 #ixzz3W6WvVVmb

Follow us: @ancientorigins on Twitter

Ancient Origins Web on Facebook

4 it is the first time that a large skeleton has been found in Eastern Europe. In 2013, the skeleton of a giant Warrior dating back to 1600 BC was found in Santa Mare, Romania. Nicknamed 'Goliath', the Warrior measures more than 2 meters high, very unusual for the time and place, when people were of small stature (approximately 1.5 meters on average).

5-Lake Delavan Wisconsin.

The warrior was buried with an impressive dagger indicating his high stature.

The news had a great echo and caused a stir, so much so that the New York Times reported the news on its pages. Perhaps, in those days, there was more freedom and less afraid of the discoveries that can change the well-established scientific beliefs based solely on theories. So, writes the columnist of the article from the New York Times published 04 de mayo 1912.

"The discovery of several human skeletons while excavated a hill in the Delavan Lake indicates that a race of men so far unknown once dwelt in southern Wisconsin. [...]. Heads, presumably these men, are much larger than the heads of any race that America lives today.

The skull seems to stretch back immediately above the basins of the eyes and nose bones protruding well above the cheekbones. The jaws appear to be long and pointy

SKELETONS DISAPPEARED ANCIENT RACE OF GIANTS THAT RULED America

CATEGORY: CIVILIZATIONS

25/12/2013

Read more: http://www.ancient-origins.net/news-history-archaeology/giant
http://www.ancient-origins.net/news-history-archaeology/giant-human-skeleton-unearthed-varna-bulgaria-002787 human-skeleton-unearthed-varna-bulgaria http://www.ancient-origins.net/news-history-archaeology/giant-human-skeleton-unearthed-varna-bulgaria-002787 002787 #ixzz3W6XDBE9E

There are discoveries that, for reasons not entirely clear, are stored in the forgetfulness of human knowledge. These findings may shed light on the distant human past; however, they are shrouded in fog with many conflicting time lines. In the case of North America, there is evidence of that these skeletons were destroyed for not reveal the origin of the races that inhabited this continent. In May 1912, a team of archaeologists from Beloit College in the United States, in an excavation carried out on Lake Delavan Wisconsin, brought to life to more than two hundred mounds effigies which were considered - as a classic example of the Woodland culture, a culture that thought prehistoric American dating from the first Millennium before Christ.

New York Times, 1912

The description of the provided skulls by the New York Times, reminds to the shape of skeletons belonging to the recently discovered in an ancient burial in Mexico with the difference that here we are dealing with more senior individuals from three meters. Who they were, and why there is no trace in the official story that we were taught at school?

These human giants lived on our planet, and, in any case, belong to humans? May this be an ancient settlement of the ancient human, survivors of the tragedy of Atlantis? Or, they were beings of other worlds that corroborate the theory of ancient astronauts? Is difficult of tell.

Archaeology alternative (AIAA) that the Smithsonian had destroyed thousands of giant humans remains during the first years of 1900. Not was taken to the light by he, the accusations are derived of it institution American of the Smithsonian that responded demanding to the organization by defamation and trying to of damage the reputation of the institution of 168 years of antiquity.

Conclusion

There is no exact appreciation of the creation, or science or theology are branches of confidence for the understanding of life, they have always played dice with the universe, human beings are puppets of the drivers, being must be the main actor in this stage of creation, take control of their powers and establish their own personal world do not let others play with their lives, have the freedom to choose, a divine gift straight man. The information is released to the public and does not flow as it should be.

The man fined God behind each door that the science opens.

Albert Einstein (1879-1955) German scientist later naturalized American.)

These data show front to science with the origin of the creation that surrounds us in the theories presented in this document as proof that the origin of humans and species in the oceans with the first evolutionary emergence of beings on Earth is based. As Clement of Rome argues, in its concepts of mystical philosophy of life of those who contemplate later development at the time of the life of Jesus, the example that emerged on how consciousness divine manifest in beings in the human realm. The same thing happens in other realms, but not has deepened in the perfection of intelligence the more imperfect innate of these creations is the human being.

127- The beginning of life

Personal disclosure based on the accumulated facts

Being an initiate into the mysteries of creation discovers flowing veil and hides eternal wisdom, hidden from the consciousness of universal creation. The universe is active mind from which everything originated.

The details will pop up in view of the intuition of this universal mind. The cosmic file of the human beings, in relation to the level of internal maturity, continuous flow of its essence to be since its appearance in this plane of creation.

The human mind is a reception between the creation and the created channel is aware of the universal mind mirror and its processes to which can harmonize with the proper knowledge. Ambiguous concepts of tradition, the adoption of myths about the processes, divert the attention and concern of deeper knowledge. Sure, someone already controls everything so it is not necessary to worry about or go beyond what is proclaimed. Through them centuries have existed, minds dedicated that have transcended it sublime of the creation, his legacy is what is studied in this history, and the countenance that not is has revealed of it hidden that belongs to a dimension of controls. The active mind by incarnations can deny that knowledge and has achieved it with the proper orientation and preparation.

The spiritual legacies of many embodied beings have brought to humanity at different times, sublime messages to awaken the conscience of the other brothers of the creation. The common mind is subject to created convictions by the free choose, are a sponge of all what is projected and touches our perception, the be discerning and accepts what his mind reasons, per their maturity. Stage use that knowledge as a point or focus of orientation, the truth of the creation is a complex world outside and inside for being, their stages of screening will cause a temporary impression of reality.

Universal laws are a spiral expanding; its truths are cosmic creations that the mind cannot cover completely the progress of data be added to the reality of each being phased. The advancement of functional reality depends on the advancement and the revelation of the eternal truths, at each level it reveals a different world, a variable printing to each individual consciousness and becomes collective by communication. The databases most of time fall on vested interests that are not willing to let the truth to flow freely. This is the dam that society must face to discern the truth of what is projected, to open individual consciousness and see the reality of life, free to think and create.

Must raise our minds to cosmic and physical phenomena laws, because creator energies, emanates from those Interior structures that not fit the genetic changes due to their structures and forms of internal change and adaptation. Cells that have kept a genetic code and are not adapted to any change, it would be another link from the nature of evolution. Laws you obey all species in all realms of creation are controlled by the emanations that invade our internal and physical environment of the energies that interact with our physical laws. What is called emotional because they are affected by the internal harmony of the conditions that establish a sense of well-being, inner and emotional that they maintain an existential balance? If we are looking for a reason for the disappearance of the dinosaurs in the forests of the Earth, due to the phenomena of asteroids and the fumes, a phenomenon that is destroyed the existence of all animals, plants, humans, seeds, eggs, worms. Possibly the effects of the isolation of energy, by entering the atmosphere block, affects glandular systems and the internal structures of the bodies that depends on them, therefore their destruction had been the lack of energetic material so that the internal laws of growth and manufacture of the elements necessary for the growth deteriorated to the level that the largest sequence of life is altered, and no longer marching in a relative period. With the publication of Pangea, where the largest number of dinosaur fossils have been discovered, alerts us that some phenomenon happened in that Time

For me is a universal legacy of the Egyptians discoveries and advancements in the neighboring territories that bring us up to the modern era with so many mysteries that clarified? We must distance ourselves from the current religions that condemn everything they cannot explain, reject it by the complexity and that cannot be internally, for being a barrier to vested interests. The Egyptians of the era of Aton were not mistaken to worship the Sun as a God of creation, because from there they developed a system of introspection and bequeathed to humanity many mysterious knowledges. Tell to the Amarna-monotheistic religions center founded by Akhenaten, Lector of the cult of the Sun which prevailed over the old religions, proclaiming the monotheism on ancient traditions; assign to multiplicity of gods their concepts of worship. Per the records of ancient civilizations which are still present as a Center for tourist expeditions. In this place is said the first system made with a concept of the creation divine and life after her death dedicated to the God solar, Aton as deity creator only. Concept this that does not resemble the concepts adopted by other religions centered human gods, Christianity, Buddhism, Islam and other known religions. The concept Egyptian recognized the divinity and the life after the death and a single deity creator, not is the same low which is have founded others religions. The unique to mature a unique concept with a firm in monotheism are the Hebrews where comes the nation of Israel and the Jews who proclaimed, which Jesus was heir of David, as the real King of the Jews not as their God.

The concept of God for the first Hebrews declares that God is the highest creator that the salvation of the human being is unique and personal, that God is so zealous in its relationship with the man who does not allow another man to intercede with for him and that salvation is solely and directly with God.

Mathematics, geometry, physical laws, the metaphysical Alchemy and all advances to have been accumulated in these times of the upper and lower Egypt, together with Greece was directed by many of the scientists, first started theories today we studied in schools: Democritus, Archimedes, of Millet, Plato, an endless list of scholars from all branches of knowledge. The fact that just as the Hebrews were founded in a higher deity and not human beings, religious ideals them distinguishes current concepts that, with so many advances worldwide, have enthroned in fractional concepts for races and gods that divide the human spirit into sects with different gods and names. If that is the privilege of humanity we must look with suspicion, look for a more universal understanding of the Almighty Creator.

Those who sought to overcome the gods in Olympus's Fables and the Roman gods not forward anything to humanity. At least those gods were temporary and present on this physical plane. Giving to beings of spiritual advancement carries the worship of gods. Ethereal with the name of a mortals, then failed to set an idea of human worship a deity disembodied as it has always been the God of all creation. Only to level world there are beings that advocate for a universal system of values and a common universal system. There are few Nations that are committed to this vision of the future. I hope not amazed with predictions that as a maximum, in future events, because everyone thinks that they do not fail.

131- Nostradamus (Accomplished)

Alert to the humanity in the year 2012

Theurgy magic and metaphysical 2012 future personal insight to create awareness of Nostradamus alerts. (October 11, 2016) Analysis and conclusions of Nostradamus: my personal version: Gaspar Pagan

This is a segment to alert humanity in the year 2012.

And today January 28-2017

another version was present when the last year of the Israel prophesies and persecutions are present at the no 70 week of the prophesy.

The powers behind the destructions of Israel will see the power of this small nation.

The third nuclear war is near to start.

Israel is the most persecuted nation in the world, the nation that was tried to erase from the earth many times, and they survive till today maintained in his legacies of the prophesy near to accomplish. Nations trying to empower of the earth territories, using the hate to call the attention and way to eliminates other etnities.

January 28, 2017

This warning was described with the purpose of raising the universal consciousness to the danger that was perceived as an outcome of the persecution of Islamists in the 1500s and the Crusades created to push Islam out of Europe's borders. The earthquakes created by these persecutions and the Templars put the world at the feet of the great empires of the Holy European Alliance against their common enemies. This alliance was founded on the principle that only the monarchies would rule in Europe, everything what not out that way would be combated and pursued.

From the bowels of the homeland of Zoroaster hangs one of the greatest threat to humanity; the hunger, the killing of civilians, the world unrest is at the gates of these generations in century 2012. A replica of the wars of the Jews, Flavius Josephus a Government closed the public truth. The people starving in the streets and sidewalks outside the fortresses fell like flies one on top of others, such as zombies aimlessly, eyes exorbitant by hunger and ulcers let see your bowels, while inside the fortress the others are wheat and food to the hungry people.

Nostradamus: 2000 to 2025 - by: Jean Charles de Font Brunes Century lll-page 87 - 3

"Mars and mercury and the attached money together around noon extreme drought: Asia fund it will be earth shaking Corinth, Ephesus then perplexity." If tests of the different dates that brings the historical relations in the writings of Robert Ambelain, add up the prophecies of Nostradamus with events that can be consolidated in this period from 2012 onwards. For Robert Ambelain "Jesus the deadly secret of the Templars and other works of authorship relates to the birth of Jesus could be 16 0 17 years before what referred to in the history of the enacted teaching textbooks to the churches. The dates are established by the calendar Gregorian. Who can prove after so many changes a date certain? If Nostradamus ignores these details, the changes were made to the calendars the possibilities that these centuries are oriented to the events. "The war (Mars, God of war; the Balkans and the Caucasus), corruption (mercury, God of thieves) and the power of money reign together; there will be towards the noon a great drought; (From money? -= the Japan live major earthquakes, Greece and Turkey will have problems.) Iran Turkey and other Islamic peoples are at the gates of proclaiming Islam; if Turkey sees an opportunity to prevail Islamism could go against West. (ESA) lll-when the defect of moles is next. One and the other does not fall short, cold, drought, dangers in borders, even where the Oracle has begun. The word Moon appoints Islam because of the Moon (Crescent) which is their symbol even in France where the Oracle has begun (homeland of Nostradamus) when the Muslims are about to commit a lack (of conception between them and the West) too large, will be known the cold and drought, even in France. I, 67 Note: Quartets that make a chain of events that intertwine and targeting Islam as the protagonist of these prophecies are consecutive. The consequences of an attack on the West by Iran and Islamic countries and possible allies, who sees an opportunity to strike a blow to Israel, Judaism and Christianity included, passing American, to be his allies national destabilize. In the present 2012 the

President of Iran is focused on these strategies. The Governments of Nations where hunger will be an epidemic by their overpopulation will in this event a chance to solve your problem of overpopulation, and as history repeats itself the holocausts of humans would be the solution to their problems. That is another nation to fend off his attackers, which cause the debacle and the extermination, a legal way to exterminate the surplus HUMAN who cannot be fed in the future.

II, 46 after large largest human gathering prepares the great engine of the centuries renewed; Rain, blood, milk, hunger, iron, and plague. In the sky, seen fire, running large spark after a great gathering of troops, another larger is prepared after the revolution (rain, blood, will put end to the good life.) Iron, the war and the epidemic, then a large comet or Fire Corps crossed the sky. Burning torch in the sky will be seen, by the end and beginning of Rhone famine, sword, afternoon planned relief Persia returns to invade Macedonia. Macedonia is a military objective of Iran by the conflicts in the European union, Turkey will see in this an opportunity to attack to Greece by their conflicts for years and at the same time will be the perfect excuse to give back to the West. On the other hand, the Orion prophecy is based on those same data, calendar officer established by religions and accepted around the world as something fixed. The end will be for the year 2015 to 16, per Robert Ambelain lunar calculations and the date of the birth of Jesus. It should be noted that Irenaeus of Lyons puts Jesus dying at the age of 51. This piece of information from someone who was present at that time shows that it is possible that manifests itself. This is a logical deduction, time to adjust the others.

A sensible solution to this global situation would be a cooperation agreement where the oil-producing

Nations released worldwide prices that food-producing Nations lowers production costs and they can at the same time doubling world food production and is supplying countries with overpopulation. Not only the United States is called to solve the problems of other powers, each country should enter in agreements for release them barriers that stop the growth of them countries producers of food. While they are used as weapons of power, stopping the supply of goods as a means of pressure to solve a global problem. The Nations will be doomed to be treated equally. Officer as a funnel as up to now where solutions fail to pass as a proper slope to the problems is solved are facing every country... Looming catastrophic consequences for all Nations, if world leaders do not depose attitudes in favor of the inhabitants of the planet, this will be his award. Gaspar Pagan (Edwin).

As a whip that refreshes the memory, history repeats itself and the beings who have not learned the lesson must confront the debacle and the destruction or return to world of peace. Everything created are looking for balance and raise green grass fire should burn the dry and the universe is renewed at each action takes where created to the same principle for pop up again a revival. To reach human ambition, the way that has been raised the hatred and persecution in times of Nostradamus, the creation of States and laws for the implementation of the Papal States and the conquests of territories in the West had the sole purpose of taking over the known world. Once you set the persecution of the Knights Templar, the rise of Pope Pius XII to the power of communism and their leader Hitler, where Franco's Spain and other allies created the power and the Second World War, financed by the power of this holy alliance of the Jesuits. The histories stop her from reality hidden for centuries. The new phase is in the city between two rivers, the persecuted will be it threatens real next. The city between two countries which watered innocent blood that ran like rivers in Europe.

136 -The mystery of acquired knowledge

The Earth stops from its bowels similar and different forms of creation. For example, a small scroll of gas that emerges from a chemical reaction contained for thousands of years in the interior at that moment leaves his cabin where he remained isolated and is combined with other sources of energy: hydrogen, phosphorous, potassium, arsenic, sodium, etc. These combinations and the internal heat of the Earth arises, stones, metals and other combinations that create a wide variety of new forms of expression when mixed with elements of cosmic emanations. An agate stone is just that: a portion of gas trapped in the interior of matter which cools and continues to grow at its capacity. It is an example of creating reactions that cools this continues to grow, enclosed in a niche that created the earth itself. It is part of the reactions that are repeated and are taken as laws to be able to explain the phenomena that occur.

The death, the change, the transition

When these faculties cease? What is this mass of energy that we went and where is going to stop? What law regulates the period between a phenomena's? - Reorganization of the matter in as many varieties of expression, every day surprise us more diversity in which manifests. The mysteries surrounding the evolution of the species for millions of years keep busy scientists, thinkers, mystics, archaeologists, paleontologists and all that branch of knowledge that seeks to unravel the evolution of species, the largest chain of human knowledge, which is the carrier of its own information and has not been able to decipher its content so far.

Our being inside, the soul and the conscience of it, when the soul is raised to the understanding of where was born and takes note of the attributes that generated it, the mind that created it assimilates its form and its gradation. That is the legacy that gives him his freedom. The pure energy of creative vibration degenerates to act on the scale of the demonstration that contains it to a greater or lesser degree. It is the way to be unveiled, act in the descending scale and take that level of existence, to interact with other grades of energies that were created by the own emanation of what was the first source that emanates, which manifests itself. The light cannot exist without the darkness. Life cannot exist without death. All forces have in its emptiness, its opposite. In the mind of the man there are all the dimensions of the creation and is it only that is da has and them performs. If we are an emanation of a creative principle that all-encompassing, this may not become aware of itself if you don't have a mirror which reflected and know auto. That is the reflection of the force that contains it so that it emerges and realizes that he cannot escape reflected in its opposite, or would lose the quality of manifest and affect consciousness, an entity of gradation that is, to do it or give to know, that reveal your hidden content. The force of creation, his passion to create ego would stop or twisted in its functions. Purity is not a world of existence because it voids itself within its own power and would only be a reality for herself. The essence that contains it and produces would revolve in own cabin and would contain her. You would not need to be known, should not emanate or share their essence with nothing, therefore, their existence would be she and nothing would make it. The laws of the universe are change and rest, everything is in constant motion. He alludes to forces of purity is back to the center of where all emanates or is projected without giving is to know. The fact that you want to catalog or give a description that generations created by the emanations of this primary force must crystallize in it, rather than evolve should return to the source that generated them in pristine nature, as they emerged from the original source. This would be the biggest dilemma that the being must face. There is no natural or spiritual law that contains this notion within the self. We can use all the powers that are aware to want this stage of regression to the primary evolution and it would not be possible, nobody would achieve it. Own power that created us lost control prompting. To be known is the ego than it creates is a need or impulse that leads to that generation of effects and manifestations in the universe, being is your vehicle. Stops in the middle of all creation, we realize, we seek to return to that source, where she herself vanishes by opposition, the halfway point, rest, neutrality, where all sets a balance. Opposing laws are attracted to create, this happens in the entire cosmic universe which generated us, the channels that separate all forces to maintain a balance. The same between planets and atoms, all materials comprising the universe we cannot grasp in our finite minds. The created realms are like the flower that languishes and continues to distribute its perfumes, scents and colors, then it becomes your seed; It contains it and returns to its beginning, returns to generate another plant with the attributes of the previous, but not all plants can be. So, it must be the wisdom of the self when outdoes itself and becomes part of it. The emanate energy of cooperation toward the source primary by an inner maturation of knowledge, own laws are a channel that gives us strength to harmonize our inner world. It is not pushing but attracting that we achieve harmony. It is not the rejection which gives us the triumph, but the acceptance, the attraction of opposites; that is the overcoming of the chaos. When unilateral forces are the weapons of power, they are a bottomless barrel, wrap everything created and destruction knows no bounds. There is no way of feeding them, consumes all the energies of the Earth and has no boundaries. They are the powerful without scruples that have managed the entire civilizations and lead them to their own destruction by the love of power. He loves the power only reveals the weaknesses of minds diseased, that not have the value of recognize is to itself same. Need the power and the domination that others suffer its weaknesses. If they can control crush humanity, so that their purposes to prevail. The positive forces of creation cannot be exhausted in any way, no power to stop them. Greed is a bottomless pit; the sufferer has been born with a worm inside and you will be consumed by it, as well as the notions of sin and guilt. Acceptance of inner demons that flooded our nature is a wrong concept that allowed be raising their spirituality and overcoming. The acceptance of that are in stages of development and each incarnation is an opportunity of overcome that condition and them squalor to which sometimes succumbs the be is the guide more sensible to redeem our maturity inside and raise our soul to a grade greater of purity, that is the path correct. All forces when an intermediate gradation achieves peace. High, low; sour, sweet; peace and war; the wide and the narrow; the right and left; life with death; the light and darkness; the movement and the rest; love and hate; the exterior and the interior. Only leading fair committed to the securities firm to defend a position of harmony at level global, with the ability to submit the barbarism that is the purpose of existing beings. Are the sephirot of the Kabbalah human concentrates in manifestation, them farms Vedic that is can study in different philosophies of life spiritual contain elements that help to mature concepts sublime of our powers of healing and divinity. From the ancient Hindu and the heirs of his teachings, Zoroaster, and many of the ancient traditions that contribute to the understanding of the spiritual life to strengthen a comprehensive concept that is not centered in one direction, to obtain a universal vision of divine power that it underlies the creation of all genres to which we have access. To achieve the strength of any emotion or factor that leads us in that direction, you need the Union of beings who agree with the same principles and purposes of overcoming the squalor that we expose the weakest. We cannot abandon us to the passions, the disharmony, that we create as beings. Grace makes the laws fairer. "But beware of violating the precepts of universal harmony, its axis could change course and everything would succumb, and human power is already wouldn't need." On the contrary, nature is a world of attractions which is never satisfied. Everything arrives at your fingertips will use it to produce something different, his passion is creating. The primary substance benefits all, nobody owns it. If someone has it would be herself and could not exist in the same plane, as it would destroy itself by being opposites in a fourth dimension, a black hole. If it were to be owned, the ego which reproduces it would not share it with anyone, would manufacture their own castles and would move away from the other mortals, as a beautiful woman. He just feels that it is their master would do his slave. It organizes everything that exists, turns it into laws of operation. If it spreads, back to organize it, create new forms of old, uses those same atoms, electrons, neutrons, photons, and every particle that crossing on its way to produce, create; harmonize the entire universe, his law is love for creation. The beauty of the shapes, the subtlety of what impresses our senses, and the senses of all creation, attracts you the perfumes, the gradation of colors that impact our admiration. If we imagine a new way, it strives to be created; her passion is to please and impress the perfection of the human being. The beauty of the forms is organized by the need to express what you want; What attracts our senses and admiration is organized per the needs of harmony in creation. In addition to the other realms that use their own laws of attraction so that their evolution is successful. He loves the desire for perfection, the union of opposites; It establishes peace in the universe, organizes its future laws, which must duplicate or change to keep its course shaped progressive, harmonious with its emanations... Being that it cooperates with these laws finds peace, inner harmony. Can verse in the plan divine a passion of exist, as the other creatures that praise the creation with their songs, perfumes and colors. Dress colors plumage to impress, attract conspecifics; the flowers produce the finest perfumes, elixirs that like to bee and insects so that they cooperate in its plan to reproduce, or create more of their kind. The fruit produced by the tree contain the best dishes to our liking, fall in love with the animals so they consume them and leave your seed on other side to reproduce. It gives the nature of fluids and ways to excite the passion and the desire to the contrary. This need stems from his passion to overcome the creation, use these smart measures since they do not have, the ability to scroll and the same need endows them with these skills. Since it originated, it organizes the universe with its main laws; It produces and creates new, to fulfill its creative mission. What nature destroys reconstructs it and repairs; It generates new forms of nature destroyed. After the blazing fire, new life arises and surpasses the preceding. In our body, it harmonizes the violations of our appetites and repairs the damage that we create with our violation of natural laws; clothes with our imagination the total of our expression, although we do not realize. It penetrates every corner; he directs his creative energies to establish harmony of the disease, which is created with the disharmony of their misuse. When violations of its laws are extreme, it punishes with rigor, destroy everything that has been created, violating the subtle laws by the freedom to choose. We amazed of what they perceive as beings, we complain about what happens to our round, own laws to punish violators. Gave us the freedom to choose and we violate the most elementary laws. If we are aware of that truth, we must search the internal improvement and enter harmony with his laws. On the nature of the cosmos they applied laws of reorganization. Our allies are the laws that create harmony and perfection. But they can be the most destructive forces if they alter. The wind with its breeze mild us takes the elixir of the life, shares its essence with all it created; same as the fire, which gives us warmth and cooperates with our way of life and existence. If your harmony is lost, it happens as a destructive thought: sweeps and destroys everything in its path. The power of thinking affects our system and our environment; so much so that the masters tell us that we are what we think. Laws of creation are harmonious with the energies of thought. These attract or away from peace and love within ourselves; You can create or destroy our own inner harmony. Opposites are part of the heritage of the imagination, the laws that we must understand to the best there is. We must get to know them and choose what is beneficial. Those who choose evil, negative, suffers its consequences. Creation is only friendly with everything what is harmonized with it, but does not differentiate when it comes to destroy.

His passion for creating never fills completely; all puts a limit to exist. Everything succumbs to make way for the new creation. All stumble to beauty, all aspire to possess it, but the ego consumes them when they become the envy of others. If you own it, you watch day and night so comes the thieves to steal your treasure. It's the balance that in your judgment on what plate putting your counterweight. If you forget, you ruin the imbalance. Learn how to master your passions and your focus in the middle of the scale; It weighs your actions, measures and tolerates others, be compassionate and friendly. Everything are pursuing the ultimate in our life, we seek to be better, happiness, true love, the exaltation of our being. We have attributes to succeed. Most searches inside and manages to rip into the walls of the time something of what is perceived as superior. But by not have a notion of what is seeks to fall in a crossroads of concepts that tie their freedom of thought, your free choose. Who discover the meaning of these internal realities must not deliver them to any mortal? It is a treasure that belongs to him. The day that deliver it, will only participate in the happiness of another and will depend on it. At the time no longer think or look for their freedom, you will only lose your dependence on that source. I know free, I know you same. Even the universal laws operate on the same principle. Others embarked on the outside search, materially, in stacking properties and things that then discarded because they do not meet them; his ego dominates them and they are victims of their possessions. When you reach the maturity of your conscience you will do herself, you Light with the light of the universe itself and do one with him and you own light, will be eternal and the matter will no longer be necessary. Within a man of light there is always light, and the world is reflected in it.

The Egyptians

The Egyptians were connoisseurs of the secret of creation where Akhenaten stablishes a superior wisdom of his race. The day is to reveal the hidden secrets of Egyptian world will realize knowledge so developed that they possessed in antiquity and awarded the Egyptians a role of recognition around the world. Pharaoh Akhenaten came to melt your material be with knowledge and clear intuition to perceive the divine causes of the human being and the highest. That same knowledge was acquired by many disciples who were contacted in the old schools. The legacy of this Pharaoh who introduced the concept of monotheism in its time was a rich School of teachings that has been treated to wipe from the Earth by those who adopted his teachings. Including their priests is revolted against him by stripping them of the authority divine that possessed. But in despair and destruction they turned off the light to the world to come. The Corpus Hermeticun, the cult of the Sun based on the Egyptian mysteries, declares Marsilio Ficino, met a new splendor. Sun embodies, in descending order, God, divine light, spiritual enlightenment and the body heat. "We are between two ovens of emanations and Dios is the emanations of the entire universe." As it is above, it is down. We have created instruments of measurement for multiple forms of energy that travel by the space; they form a spectrum of all the elements that move to Earth. Men of all branches of knowledge, observers and mystical looking for the way to give knowledge to the humanity of the laws that affects them. This creation is directed at forces that we perceive with our five senses materials, will have shown the wonders of those measurement systems. Musical scales give us another notion of these vibrations that we can create in our physical plane that we can play with knowledge and personal domain. Double vibration for use as a harmonic manifestation of our perception of the cosmic scales to which we have access: light, energy, emissions, solar and, beyond that, the projections of cosmic vibes that reach our planet. Scientists have overlooked a simple law of nature, which is the creator of all these manifestations and the need, the passion of creation. In the entire universe, if something arises and manifests it is because there is an urgency of some kind in the own universe for that matter will regroup in what manifests physically. We have come to the domain and the knowledge of the more infinite division of what is the matter: The Atom with all its components. Cohesion of all parts by laws of attraction, the separation by repulsion and a third law joins them, characterized by harmony there is neutrality, vacuum, the halfway point of meeting of any power of cosmic harmony, manifesting a product that interact. So, these end as are manifest should exist, in the same vibration of matter, an energy superior to all them that les da presence in the universe. Being negative and positive, as we have listed them, they must obey to a third force that compels them or harmonize them to manifest and create. Imagination brings together these energies and shapes them until they materialize on the physical plane. This is a force that exists only in the divine mind and the being and nature itself to universal cosmic level. Internal momentum that should be full of energy pulses of the quality that is and that harmonizes within our subconscious mind and then stop into the conscious phase, where the momentum of developing that creation fills us with a great sense of satisfaction. So, great is that it moves a world economy at all levels that the human being can imagine. We must bear in mind that when one speaks of being in sense of expression Dios must understand father-mother, man-woman, acting together, and grandfather and grandmother emerge as attributes of God himself. They are elements of the perception of the ancient cultures that were closer to the true manifestation of creation, which has been lost in the clutter of the interior space.

It's a natural intelligence that sorts everything that exists and operates from its form more simple or complicated expression, that being part of what we are we can perceive it even forming part of us and of the universe, because we are she and we project energy toward the source that provides us with features appearing physically in the material. We use a subtle force of attraction, a magnet or static strength so that energy is present in this plane and aroused for periods of time. There should be a vacuum of this somewhere in the universe, with all attributes, who manages by attraction to descend toward the material and spiritual reaction that then exist for periods of manifestation of creation in this plane should return to its original shape. The energies of the material are incorporated into the material realm and the soul, to the sublime field of the soul of creation.

146 -The maturity of the soul

It has been since the beginning of time knowledge that only few have access. A secret of thousands of ways describes dimly by those who managed to imbue this link in the soul with the consciousness of the creator father. Known with certainty that it is a place where the soul ascends and is climbing through a tunnel of light and down back turns into a dark tunnel. While traveling back darker it becomes. This experience has many human beings who have gone through a process of personal resurrection. Personally, know a humble person that spent by that experience and of that form knows, many thousands of people have had them same experiences that not manage to explain and that saved in it more recondite of your soul, because the misunderstanding them scares and not is dare to declare.

The maturity of the soul is the experience of evolution.

Light:

Analyze an issue like this would be to ask where to go light when it is turned off. A power switch is the mechanism that separates one reaction of the other. We only got that its temporal manifestation stops flowing, but power remains latent until the switch is operated again. No matter the times that it is done, the energy will manifest itself. It will be screened at the capacity for which it has been regulated. We have an example that compares to this phenomenon in the bulb. Once the bulb life reaches, its mass of material must be replaced by another. In that period that takes change, energy remains latent until a new body takes the presence of the former. The creation is a cluster of infinite energy can be taken as an example this electricity and which has the ability that it can manifest as many creatures without limiting its capacity to manifest itself. It same in the capacity electric in a single source is can connect thousands of light bulbs. These are laws that man can control and channel with their knowledge, but cannot create nor destroy energy; only because it occurs and you will be useful. It is a clear comparison they have a better idea of what is the creation of beings. The statements of spiritual laws are derived from my personal experiences in knowledge of mystical laws acquired as a student ("AMORC") ancient and mystical order of the rose and the cross.

The institution exists at world level as school of initiation, which spread a personal knowledge, these issues of such profound dating from about 3500 years. When the human being of creation, which can express the Kingdom of the father inside, known that his destiny is to project the divine spiritual heritage which will accompany it during its period of existence in this earthly plane, that the divine plan of God who created us. Once it transpires that material body will again and will be diluted in a scroll of power that will regenerate until it is drawn back to this earthly plane.

The Mystic that knows the laws of creation knows that once the soul detaches from the body, the darkness, which are always present, take the place of the inner light and matter is projected in another dimension, as well as the vibrations of what were the attributes of the soul. The mystery of the integration of subtle energies that emerge from the body or the soul does not lose its qualities or fade as it is the common belief. There is a level of containment or dimension of these energies that have access to the soul-personality. You call heaven, cosmos or the name assigned to it for the deviated from natural processes.

Beings that are prepared in this regard can consciously enter this area or dimension of divine creation and inhabiting by periods of regeneration to return to this level of expression as a new living being, a new child with a new soul, a new living being.

Reincarnation

Reincarnation, that is the legacy of the creator father and the teachings of the Master Jesus announced in the Gospels of Thomas and Philip. Of this act arises a subtler vibratory rate that raises the vibration of this corporeal mass; There are the laws of attraction and repulsion, and the magnetism in the be. This is what keeps the internal universe in continuous movement, emanating at the same time new manifestations which occupy the spaces and create your environment of existence. If someone imagines the other forms that are manifested in different species, each vibrates to its ability to attract energies that complement their expression form of the primary source. Only human beings can perform a higher world where the divine laws of God are in harmony with your soul-conscience. The attraction and the repulsion are two qualities that Act and a third force is manifest of the harmony between those two forces counter and emerges what is called the love. So, understood, it is a law that governs energy where there is a neutrality that harmonizes the forces. Harmony or balance of these is what we call love or marriage with each other, without mixing completely. The force that bring all the energies, into a small universe as in the great universe above all, this manifested itself a force that vibrated as an emanation arising as a scroll or - glow or emanation - cosmic sphere of forces that interact with the previous ones. Is there where, by interact in these basic laws of creation, arises a unique and true dimension of being. Mention being is the paradox of the creation, the emanation of God.

It be a scroll three-dimensional of vibrations is groups, involving energies of matter primitive in the be. Meet empty content matter, they make a combination that gives them a definite strength (Cohesion) and sets in motion the other materials. This combination is a dependency. The kernel that is created will continue to exist if you have access to these cosmic energies. Is like a fourth dimension, where empathy creates the field where being cannot penetrate; only his skills emerge and share without mixing a radiation of intelligence so that it can interact with them without being able to draw it out in our dimension of known manifestation. An indescribable emotion that seems to cover everything we projected and we cannot penetrate, but shows us and perceive its existence and per what you Arqímides said, God geometrize.

Brain components of their atoms are those who come into this small universe of physical reactions. These mergers are the mist of subtle matter that gives the material properties to everything that exists. What we call the conscious mind in being is the attribute of that energy. Gives us the ability to realize the movement, the dimension's time and space, an intuitive knowledge of its contents, because we are her own. Emerges a dimension in any part of the creation, either within us or outside us, where these combinations are drawn after created. A cosmic dictionary of our relationships and emotions. We assume a control to express emotions, and describe them as laws of principle. Of the same form that is maintains this reaction and combination of energies when the being leaves of vibrate in your body material. These are returned to the cosmic part where emerged, and the richness of emotional maturity is part of the soul or substance universal, (Akashic) files. Internal mansions of the understanding of the surrounding world to physical expression, is the cluster of energy soul spirit added energies purify themselves for new experiences and the benefits of what is pure and divine maturation. The repetition of these States in overcoming is what guides us eternal life, because they are these energies that continuously manifest in new creatures and returns to the cosmic consciousness, which is God. Only the verb we use in our spoken form gives you a sense of understanding and action that surrounds your content in our minds. Into this vacuum, the conditions are created in our bodies to produce the sounds appropriate to give a sense of expression over time than we can imagine. Those who first discovered this communicate them to others in spoken form. Due take thousands of years, form is a vocabulary that were cataloguing to be able to communicate is. The myth of the cave is an ancient example that describes how the imagination of a human being is unique; It has a richness that you can share, recreate, and projecting the same images of our other minds that are not aware of the phenomena. The lighting creates mirages, which see through them you will discover the wizard and you will be warned of the danger, in advance, will be grateful for these assertions that no one will be able to capture.

151- Lighting of the self

Being that it will harmonize with its light on this earthly plane will have the blessing of Intuit and don't need to see it, because it will be part of it. If its light reaches our interior and is projected in our consciousness, being projected on a space-time tunnel, where it becomes the portal where your light emanates the presence, the portal of light. The soul, which is the product of this interact where emotions more subtle and comprehensive primary cause staying temporarily belongs to us and lets us enjoy its surroundings and realize its depth of happiness that we have the right to aspire.

Today, with technological advances, history that has been designed to us through the centuries is constantly changing as we find new discoveries. The evidence discovered change each time some of the realities shown so far. The story must be updated as many data taken as certain that we should examine very carefully the text and start again. This has happened before and is repeated by cycles. So much energy devoted to create intellectually a world to save experiences for posterity and turns out we're floating in the air and we know many times is collapsing before our eyes. We realize that what is taught is plain content, trumping us knowledge without sense that the teachers teach what they learned or what are given as education. Keep our inner world crushed by confusion created by concepts of repression sometimes pushes us to see ghosts in the worldly things that have no influence on our spiritual realities. Being that it is influenced by these teachings that degraded its condition of being despite centuries of social dislocate, have emerged figures worldwide have exalted by his convictions in his ideals and have guided crowds have a belief in something that feed them to continue the search for the truth. They have fought on the side of the weak and the poor; have been sitting Chair delivery and humility. They have been enlightened that have been expressed in favor of the ideals that they have accepted as their North and they have fulfilled their mission. They are worthy of admiration and have transcended history as wise and humble in their convictions. The less prepared in school, but with a commitment to the truth and the struggle for human excellence should not surrender. They must give account involving global humanitarian leaders who give the daily struggle for humanity is exceeded.

Intuition

Statements by the creation of the self and everything that exists in the universe will be ever captured fully by the finite mind of man. Can Intuit a vague notion of what could be last in those moments. There is a book of history that we educate these issues; with the only thing that we have is with writings that they tell us the stories of characters who were involved in the events, and which have been documented to give an example of life in those times. Analyzing this aspect of human knowledge in the field of metaphysics is as close to what has been reached. The stages of knowledge that have been cemented and accumulated a history being met personally of the rarities that manifested themselves during these early periods were being projected extinct and unknown qualities for which we inhabit today. Those more profound minds have dedicated efforts mental and physical to even approach is to a stage rational of something that seems logical. For more descriptions that can be written, it is a knowledge that is to the mind and the conscience of those who describe it. The great thinkers have left us knowledge that we need to update and continue the search with these databases. Looking in the files of my memory find that these were attributes of the mind that not needed of preparation. The attributes of intuition are based on aspects that being mature in times where only hung to capture any phenomenon to happen beforehand. It was a magic for many who were not trained to live in places away from these populations. When they were with these reactions they were surprised to see that they said things that were to happen. Hence begin the magic, divination, because they looked at how people were amazed of these powers. See that it was a condition that could be exploited to impress and charge for these tricks, became fashionable charge to demonstrate phenomena that the mind could not explain. Today we see the Magi and the compresses creating tricks to please the mind of those who love the magic. True intuition is not paid for those shows. This was an attribute of the precognition of people away from civilization and he used this ability to orient themselves in their daily live without realizing that you were doing amazing things for them was a routine.

Destruction of the original knowledge

Much lost by manipulation and the interest created. When the first consciousness manifested and interacts in what we associate with a Creator and all the qualities that each be manifested or perceived, and awarded on a scale in understanding its functions per the race and the values that practice it. They are the possible imposition of any leader in any stage of domination and similar concepts of that culture. From the most primitive species to being more developed everything that occurs in the future they will be attributes and properties of the same universal mind; its laws are the basis for understanding life on Earth.

The legacy of the beings that were devoted to documenting the ancient history is another flow for us. From the time of the Egyptians, the Sundial, the construction of the pyramids, the equinoxes, mathematical statements, heliocentric theory of Aristarchus 230 b. c., and all the knowledge gathered in the large schools of Alexandria in ancient times. With the destruction of the library of Alexandria, where the most extensive knowledge of the most forward-thinking minds of mankind, scientific discoveries, were catalogued historical, mathematical, philosophical, and all branches of knowledge, (many) have been lost. Do not know if were stolen or hidden, or truly burned. When this knowledge was discovered by the Christian Church and saw the growing threat that hung over her, she quickly sent a mob of zealots to raze everything to realize the mistakes that put into practice, trying to formulate the deification of Jesus. Sun God, incarnate in person, by Saul of Tarsus that think a God for the Romans and its intention to take over the rule of Nero. He having doubts that you seek and you will find. The important part is that mental disorientation and the approach of reciprocal power of beings that react to the arguments focused on erratic culture, misaligned forces and directs them to concepts that form a bubble of fanciful creations. From that source is continuing to feed the imagination of the thinking beings, minds started misbehaving distorted paths. What the mind transforms into realities of the divine energy and its variations coagulates into a reverb, this will continue to grow to the level that the reality that is updated each day will be absorbed by most those who return to a new stage of maturation. Children entering a turbulent society by their erratic conceptions, which have matured that civilization, will disappear. The legacy manifested them in civil orientation loss. The repetition of concepts unclear completely directs imagination to create Interior forces that eventually lead to uncertainty. It has awakened it from the dullness of trends created tamper. That must be the awakening for future civilization. The manipulation of genetics, the subtle cosmic energy, and the emission of energy from outer space, manipulated by the systems created for the welfare of human beings, destabilizing forces that dominate and control the balance of creation. We are treading on the Bank to control and manipulate energy that did succumb on many occasions the very nature of the planet. The imbalance of forces we would lead to what would be something like the Atlantis and Lemuria. The common man has not matured one way to harmonize with the energies include balancing self-thinking with the variations of the natural laws. Uneasiness, fear, failure to obtain assurance of the entities that promise you to protect him, the manipulation of information by chains of transmission of information that do not have a control or Veda of sometimes make statements without basis, resulting in unprecedented cataclysm. If I mention here of these social elements, the fear of death is also another incentive for such attitudes. Therefore, where these declarations are in this book is to alert citizens that, if there is any change in the world to the living conditions, will not be the end of the world nor much less the end of human beings. You make a statement like that will be missing the truth. No one has that power. Only the eternal Creator enables us to discern what form must act and not be afraid of the threats.

By the statements of Patrick Gerril, author of the book "The Orion prophecy" these phenomena have been before and we are still here. Their statements are accurate and credible until the time proven otherwise.

The important role of France and Spain, The holly alliance, the second war. The colonization of Americas issuing special powers to inquisitions process, degrade the human conditions to a slavery. An important population of 70 million peoples were demolish to a 3 million. That was the main reason to start importing Africans labors to Americas.

156-Inner heritage

The inherent qualities of each separate manifestation to transform to the needs of development and the obstacles that must be overcome. For a lizard, for mention an example, your development you require train is of media engines for save obstacles, a tail to keep a balance of movement, properties in their legs for Ascend on trees and a countless of features that have evolved to its heritage and needs of survival. All beings respond to the same needs and are adaptation evolution that revolves about their needs and the region where to develop. Plants save a genetic code of all the attributes of manifestation and duplicate them in their seeds, and thus continues its survival. Everything is otherwise creating a disharmony in the Act of creation and that's where manifested disease, the disharmony between the matter and the laws to be followed its course. Symbols with one or two snakes seen coiled in what represents our vertebral column where are screwed up our brain show an expression that both parties counter of our body and their contrary powers gather in the center of the brain, and if the disease has attacked the one to the other, and then the internal destruction of diseases. The Egyptians, Greeks and Romans ruled this knowledge and left a great legacy science and medicine to humanity. The bases have been created in terms of the material heritage of being. Science must deepen diseases since the beginnings of the evolution of the self to focus your search to restore health. Earth is disorienting for stages of magnetic gravitational energy, itself shifts as well as animal species. For some reason the same way to operate the human body this not orients itself by itself, that is why the ancients had that information domain and practiced concentration exercises and internal orientation of their material bodies to restore health and harmony with the cosmic forces. It was a virtue for them that spiritual legacy which makes them beings with superior knowledge. The material body evolves and creates methods to combat diseases and evolve to stronger forms to avoid the disease internally. I've mentioned it before, and it is that once any kind find a way to modify its structure of survival it will evolve to a superior in demonstration species; even from the first creatures have developed other species, resulting in other different and other qualities. The first creatures have developed other species, resulting in other different and other qualities. In their original environment where all forms emerged and varieties that they shared the same facilities of water, and which by their nature evolved under surrounding conditions with the other, and it an ovulatory and it expels outside its fluid playback, where there were no dominant laws of procreation, since the DNA of evolution were organizing in the early stages of development. The fact that these subjects are combined without control is the basis of the diversity of creatures that populated the first manifestations of life. Climate change and storms have influenced so that through time life to disperse in the oceans and all the territories of the planet, from which emanate the qualities that have enabled us to an evolution than other creatures. Glands which emerged from the combination of sodium and potassium was created where the quality of energy attraction and fanning the first cell that catered for the proteins to its own expression endowed her capabilities to build a means of development and evolution of their primary skills attract different conditions on the composition of the elements at its disposal. While the division of cells additional it fanning along with them powers that is were gathering, with the form of generate new and varied energies. Shake with neurons created from this first vehicle of expression and combined with other elements that occur together in the same room and so on more complex forms. This Center as gland provided primary neurons that shared the sunlight that gave them the necessary elements and force of expression to generate that after a long period became human beings, endowed with a heritage different to other species that were created in unison with the self.

The road

If the human being will be the potential which has in its interior and the freedoms it has inherited it would account that has access to the Kingdom of light, and dwell in him and return him to life, and return to the Kingdom in his departure and change of substance of life and death. His soul does not go away; only some changes of expression. A State of reflection to another and final integration in purity to its destination, which is the sanctification and communion with the divine substance that received my verb enters the source of my being, because I receive me as knowledge and put it into practice in your own be. Once between in communion with the nature of the father, all you will be revealed; exposure to its essence in his healing. The soul - spirit - is the part that is associated with human emotions, which at the time of starting and abandoning the material body is integrated into the nature of God: love, compassion, mercy, etc. The laws of creation taking mental stage to give way to understanding a little more definite and elevated processes that come into play in the maturation, and define laws and principles that make a true definition of internal processes which dominate the creation. Those who do not have a comprehensive knowledge, perception of space-time and all dimensions that being attributed to what they perceived as manifestations of what they call God. The inner stillness, the halfway point where there is no conflict of any kind, the abandonment of ego, self, being in the middle and share the flow of everything that manifests and is perceived internally. Nor is music, nor is water, neither is river, nor spring; It is when it is man, or is woman: it is one. The consciousness of God has no sex; This is a dimension of the attributes of the human mind. To attract the divine substance of God should prostrate in the Centre of the self and leave the silence and deep peace. Where is abandoned the self to make way for top soul that permeates the universe, where cease domain fights. It is integrated into that dimension where everything is one thing and the created light merges with the material. Up to that level of understanding the teachings of Jesus, rise as it was endowed with the understanding of the nature of the father.

Universal mind

What unites us as beings in creation and covers total dissemination of what we are as beings that we are aware of our existence as living entities. Universal mind is related to our behavior and our inheritance of own development as individuals at the same time is part of a collective mind which is intrinsically linked to our individual being and the time unites us to all. Each cell of our being, as well as every cell in the species, it has both collective and individual intelligence laws that dominate it and responds to them by impulses be harmonious with its surroundings. Our body has a universe of different functions; the characteristics of each are different from other. What is going to be a skin cell is going to manifest itself as hair is and so on, completing a chain or cycle. This in turn forms an interior universe, which in turn is linked to what is the total manifestation or human being. The attributes of being are the expression of a world that has traveled to this plane of manifestation to be known...

The inner master

He is the one who has managed to overcome the passions, the ego, and the understanding of other beings and weaknesses and suffering. It has come to overcome the initiations of the life and his consciousness projected him as a teacher. Also, is that has last the test of suffering all the passions and know them emotions human. Overcoming the ego is the Mission of the inner master. If you declare that the thalamus gland and cerebral system are those who keep these correlations and dominate these manifestations which we attribute as thinking beings, we should assume that the master in the human being would be the domain and control of these areas of emotions and passions that we have linked to our internal development. In this earthly plane, many are those who want to take control of this freedom of choice that we inherited from our Creator. We must admit that a high percentage of beings do not possess these qualities of inner maturation and depend on others to address them. From the most remote civilizations have arisen teachers who are dedicated to guide to the humblest. But they were not so selfish as to seize their freedom to be themselves.

Agatha Stone the inner altar

When intuition is an attribute outside of our being

It is a different world where human life does not exist. Is a simple stone that grows from his inside towards the outside with a hard shell that not allows guess its content and continues growing by years and years? It follows the laws of gravity, but what distinguishes it is that, despite being influenced by the laws of gravity, it expands outwards like a small universe. Generates and creates new forms, at its Center are the circular lines, curves, and a range of colors that do not have a defined pattern. The demonstration of their interior qualities gives rise to everything that emanates and is discovered inside. It reveals a process that should be like what created the own universe and a combination of gases interacting with the energies of creation have generated what today catalog of universe. In appearance, not follows the quantum laws that dominate the universe outside, because she is own universe. Fumes that were trapped inside originated a new way of manifestation. New for those characters that marveled from antiquity to the present by the variety of its interior, per the amount of accumulated gases and formed in a bubble that determined their colors. For me it represents one of the qualities of the own universe to be generated for the first time. Gases of the first sneeze of the universe and the combination of energies that sprang from those first emanations arise given life to a world that expands with the repetition of a primary law, the clothe all with layers of maturation to cool and protect its internal content, and that it occurred outside inward. Once temperatures fell, the vapors are condensed and remained trapped inside and began the real creation of the universe in miniature. The rest is history. It vibrates only to own will and capacity of existence beyond the reach of many expressions that open spent as another element of evolution. Large men are having excited with its rare behavior, their colors bright, the mystical vibrations that is you attributed. Connoisseurs in the antiquity attributed a meaning where there are engravings of civilizations oldest knowledge. It emerged from the depths of the Earth; It was born of its own will emerge, like the universe itself. The human being discovered and used it to make known the history of humanity in its principles. They rejoiced having an element where they could preserve symbols to express knowledge and curiosity of the challenge that was the invention of the first inhabitants of the planet. It has been transmitted to us a power of the human mind in evolution, know the knowledge so that new settlers produce more complex systems. Those who dominated the deeper knowledge were those who gave us the story, worried exalting the relevant details of events and knowledge, for future just like the universe. Phenomena covered by nature exalt us and concern us today as our ancestors, and they were closer to the truth of things, demonstrating to her around and we inherited their knowledge of its allegories. In the construction of Solomon's temple, is one of the stones that the architects, rejected as this is part of the material that was used to build the first temple, where symbols of the temple and the sacred teachings were recorded, and mistakenly left abandoned that had recorded the most important teachings. That tells the story. "Show me the stone that the builders rejected; this is the cornerstone." He was worshipped by the ancient, who considered her the stone of science. Applied in healing remedies in Egypt, it was used in the pupils of the eyes in the images of the gods. It was used in the prints, jewelry, and famous buildings, such as the Taj Mahal, in the India.

Inside the stone,

In them traditions, is used to cure, as absorbs them energies negative, that should be its source of growth, as occurred in the first stage of the universe; the duality to the mix creates the Trinity and the second law of creation, from which emanates the third law of physics. A stone can absorb negative energy and combine it with other energies inside. In this combination of energies is like a recorder's information, as if it was a computer, and you can record intelligence of all energies in nature and energies that entering the Earth and out of the Earth. It is like an encyclopedia of energy, and if you would discover in it the key to interpret that knowledge, we will get part of the history of the universe. The pieces that are used to store knowledge in computer chips represent a challenge for the modern inventors of technology. This can be another key of the intelligence without discovering. This is the story of the stones for the human being. It reflects the cosmic in the planet, and if the knowledge recorded in the stones could be rescued, we would have a more comprehensive knowledge of humanity. The history of ancient peoples surprises modern man by many of their stories, which occur for long periods of experimentation of phenomena that are incomprehensible to our scientific advances. The ancients left this knowledge, but its keys have been lost understanding of being common. I, observing the behavior and reactions of the body to stimuli of protection against invasion of bacteria and viruses that live in the environment, particularly, recreated a system to protect the body from infections. Note how our system is trained to protect is of a form internal and react with an intelligence higher. A negative energy source emerges from the outer space and reacts with the material contained in the runway of a certain area of land and produces these reactions. Regroup these is what comes from inside as the stones. Of this law the father gave us the legacy for our protection, as an agate stone

164- Alchemy (physical and mental)

Science hidden for centuries in the gibberish that the wise ancients invented to save his knowledge, since it tasted like something bigger and that only they could understand and control, and bound them to a unique circle or club. The sea of knowledge that has been lost, since veiled keys and symbols created many were not transmitted to anyone. But, even so, and although scorned by scientists, laid the groundwork for modern chemistry. Not so for many Mystics. The ability of transform internally it part remarkable in attributes more sublime that paid flashes of lighting to our soul by means of the combination of energies higher and the understanding of cover the nature divine in us be inside. Only in the archives Akashic is recorded, the heads that rolled into the different struggles for supremacy, the ignorance of superior beings, the history Neurons, had no idea how God exerts its laws. The information of all the events at the human mind and the aura of the soul are returned to a cockpit of the universal mind. It is difficult to extract this information, but the only one who can imbue that domestic legacy is the human being. The human mind can transcend the realm of creation. Many are those who have managed to transcend that limit of the cosmic mind of the human consciousness. Those who have been blessed tribute to humanity. Deep piece of the universe in their souls, lighting and discerning, some hidden world of energies of vibration that is, the truth will be revealed, I hope that humans can believe and receive this as a gift of love from the same creator and their laws, enacted by the beloved disciple son of God as every created is also the son of God.

Handling these subtle forces under knowledge was the goal of his research, decorating with grandiloquent names the qualities and processes discovered by them. In this symbolism, they concealed their findings and sealed them as knowledge arcana. He spoke of the formulas to create gold from other metals, following the processes of the own discovered laws of the nature of the creation by them. Many say that celebrities could decipher the mystery and obfuscated to the world. One of the main reasons for these attitudes in hieroglyphs and veiled formulas express discovered knowledge is because it was the only way to preserve the purity and the authority of what is discovered.

On the other hand, institutions that showed power did not tolerate freedom of thought and only them by Decree revealed and imposed what the people should know. They only authorized rules that dictated what the people would have access, that mode is squeezed ignorance. Other search engines were more interested in the Alchemy of the faculties of the self, the evolution of its capabilities to higher levels, the transmutation of being inside, bringing their qualities to a State superior understanding of the events and functions that are revealed through our internal contemplation. It uses the introspection to do is conscious of the world that evolves in our interior, that obeys to them laws of the same universe outside. Search how to decipher internal processes to create.

When it is accomplished, be it harmonizes with them.

The transmutation

Use in transmutation to will on top of manifestation energy to the soul and the spirit can vibrate at high and subtle scales. In the long run, these meditations and knowledge awareness rises and harmonizes our self with the divine emanations of creation. The true teachings of Jesus to the Apostles made emphasis on the correct way to prepare the body, so that it became subtler in its earthly form, and thus to receive the emanations of God with greater force. Being should direct its efforts during its existence on this earth plane becomes more aware of the divine nature of God and can assimilate with him. Power principles taught to citizens and followers of Jesus in his time give a clear idea of how to prepare your body based on vegetarian food. The rituals of daily to prepare your body to a subtle form, knowledge that will void all mental creations of humans themselves if it reveals them the true nature which affected the knowledge accumulated for centuries. Since the time of the Egyptians established a knowledge that went from generation to generation and that was maturing in the minds of citizens and Kings or Pharaohs of that time, with gradual changes and a maturation of the knowledge that was shared with different civilizations that met the world and its history. That humanity does not be surprised when they begin to disclose these findings that will soon come to light. Here outlines only a ray of light between them.

The position of the statues with feet apart and her hands on her thighs was another way that these energies were concentrated on his body and the negative not to blend in with the positive. All these skills were utilized in their practices of resurrection or astral travel, which dominated and showed in their daily lives. Encompass all these phenomena that were used were to them a knowledge so great that they kept it as the greatest of secrets. Aligning themselves with electro-magnetic energies of the planets and stars was another way to maintain a healthy and harmonious life. Only priests, priestesses, and those servers of the King were instructed in that knowledge. There was always a group of these so that knowledge will not be lost. The sacred writings were maintained under the control of the Pharaoh. This knowledge and other secrets that were obtained from the ancient Egyptians combined with the Hebrews, Essenes who helped train some priests in the hermetic Kabbalistic secrets when they were in captivity in different periods. Domination through knowledge of the arcana of the body at that then and now in day is scientific knowledge tested by Mystics who engage in these studies. Hereby they arrive to set up an Egyptian cabal that is closely linked to the Hebrew and we cannot distinguish what is true. What today is known as Kabbalah and is used to provide a knowledge vacuum, and just to prove the deception he can be subjected to human beings is not covering this knowledge, is something deeper and involves immutable universal law. The mystical Kabbalah has key numerical and written in symbols that can be translated by a connoisseur of the hermetic symbols. The fact that this is formerly kept in this format does not mean that that is the key. The ancient temple of Solomon was saved prints the symbols of the true meaning. These were lost by the destruction of the temple. There are many people prepared in the knowledge of Kabbalah that are close to a true interpretation. But as all the hidden mysteries, there are no words in the vocabulary of the languages which can give a real description of its meaning and create a mental image that is equal to its mystical meaning. By each pasteurization that penetrate this knowledge and remove an interpretation of what can access will be a version personal that never is like the concept of another human being. The Egyptians dominated this knowledge. Despite the errors in some of his views, gave many humans who came to be initiated into secret schools, including many Greek and Egyptian citizens, not to mention the area of Alexandria and an outstanding disciple of name Jesus. This information may substantiate when it attributed part of his knowledge to the Egyptian magic, as well as the disciples and other mages mentioned part of the story. The search engine will look like something unusual, but if you look in the light of current knowledge was knowledge of universal laws that today are the medicine and modern science, applied in a time where the preparation not dominated such knowledge. Even the early Church condemned science for that ignorance. To go directly to the subject of his discovery in ancient civilizations used a method of transcendental meditation. The method was called (Mer - energy) (Ka - body) (Ba - soul), if they are looking for what the Emerald table from the beginning of time will have an acceptable description of what this means. The principles behind ancient rite are to align with the electro-magnetic energies of the Earth. The ancients had that knowledge and practiced it daily or for periods of worship to the higher laws of the divine nature. Dr. Kristen Neiling explains that its discovery is documented in many of the practices of ancient civilizations. Your information revolves around the alignment of planetary electromagnetic fields and their influence on human behavior. It is known that it was current about ten or began a variation in the alignment of electromagnetic fields. She explains that it is part of the changes in behavior of human beings. Their States of excitations, irritability, the overpopulated, emotional, etc. is also related with the changes in the solar atmosphere that already has begun to feel in this year 2012. She explains (Mer) means opposite rotation energy fields in the same place. The body Ka and Ba soul. Logic is set because the Earth is rearranged to change and the animals do it, but the human being does not. The Egyptians and other races were imbued with a knowledge that covered our inner creation. Territorial and cultural changes turned off that pace of expression of a notion that was and is the basis of the spiritual growth of humans. While you were erasing those notions is turning the legacy and the advancement spiritual of them beings created. Expressed in old prints is one of the principles of that knowledge. The input and output of the soul to the human body, Bird BA-(soul); Ka spirit or body) the two principles in the body were moving energy (Mer) in this earthly plane. The body of foot means (Ka) body, the bird, the figure with human head and body of bird - at this stage the Ka leaves the body towards the underworld, then down the same being or soul returns to the Ka where the dark Sun represents the return of the BA - Ka and arises the Mer-energy in the lively body returning to life. Is an engraving that recreates the principle of reincarnation? The rites of this ancient legacy are exposed in the study of the DRA. Kristein Neiling, in a study for NASA. Later, I give a summary of that principle - one of the keys to the inner harmony in being. We concentrate on life in our demonstration centers external or materials. The most unknown knowledge comprising our inner creation, changes territorial and cultural went out that rhythm of expression of a notion that was and is the basis of the growth spiritual of the race human. While these notions were, erasing was turning the legacy and advancement. The Egyptians and other races were imbued with knowledge and spiritual consciousness of created beings.

169- The initiates

The art of awakening a universal awareness of the knowledge born of a preparation which is carried out by teachers who guide the searchers by a path of learning interior as tune with the energies above, if so it can be called. Being awake inner intuition knowledge or condition that gives it a perception of realities and veiled truths that have gone unnoticed by many people since they do not have that sense of orientation. Everything write is usually springs from some unforeseen source. It's like knowing the inside of things, a mode of inspiration but with a certain knowing of what is written but some source of starting to take. When leaders are those who provide, power is not easy. At the time of the Foundation of the largest empire on Earth is not less true that the architects of its foundation had knowledge of what they did. A King needed to unite opposite factions for good or bad and races would not enter contemplations about which the reason belonged. Only the fact of solving problems of State was sufficient to accept a faction that was random with their interests. 1-the Templar in the middle ages disappear with its fleet of boats and cease them attacks to the Vatican and their allied kings. They disappear from Europe and travel across the Atlantic and he is established what was known as pirates and privateers for a period, because they were being persecuted by enemies and exploiting its advantage of sailors to strip opponents of riches and your travel charges. It was later located in the Americas, and thus formed the beginning of what would become the great nation of the United States. Wonder what the real mission of Cristóbal Colón in his trips to America. We are told in school that came looking for a new route to the India, in search of spices and it is not known how many things were invented not to reveal or true purpose, to emerge when the route of silk and spices were already known and was made by land. What spices would be so coveted to venture is to them waters unknown of an ocean that said not know? The story is always told to leave a notion of the truth intended that he is created. The scholars found that she already knew of the existence of these continents and were listed on maps as far as in the year 1400 or earlier. It is known that Cristopher Colón had a copy of these maps. Where do they come about? It is possible that they were acquaintances of the Mosarabs who invaded Spain, until about the departure of Columbus on his first voyage, after the Arabs leave the territory? I venture to conclude, even if I'm wrong, that Cristóbal Colón had some command of the Catholic Kings of Spain by common agreement with the Vatican to locate the fleet of ships Templar who survived the extermination of its forces. While Cristóbal Colón traveled to the South, they were in the North, in the Atlantic as pirates and Corsairs among groups who settled in the American continent territory which afterwards were the first colonies of the American nation. It shows that they swarmed in the seas and the northern territories until Cristóbal Colón came to the West Indies. In fact, was his discovery in the year 1492, if you search for the story will be that already to 1513 was founded the city of Saint Augustine in Florida by Spanish settlers because of these trips, and in fact the first city on the North American continent... The other part of the Knights Templar was like pirates and Corsairs in the islands of the Caribbean Sea.

Our interior

The processes of thought, imagination, human emotions, joy, compassion and other complex processes that tie us to the creative forces of the cosmic universe, all that energy that comprises the vibrations that arise from our being integrated to the universal part of the cosmos and creating what Mystics call the Akashic archives. If humans had evolved to a spiritual level that is assumed is the purpose of our presence in this physical earthly plane as a living reincarnation of our evolution towards spiritual persecution of a being superior to the Supreme spiritual elevation. Our interior is home to stored memories of past lives that moment comes to our conscious mind. Us reveal another knowledge alien to the reality physical in which we find in this flat of manifestation and possess an attribute of discern truths that for many are unnoticed. He sees our inner body and perceives its functions give us knowledge of the wonders that happen in our being. All functions of us be happen with incredible accuracy, if contacts to us voluntarily handle them don't we movement long live. The laws cosmic that regulate our development spiritual is complemented with which regulate our environment material. The harmony between the two is the goal to maintain. The set of cells that make up our bodies have their own system of storage of data of each function; they are in communication through our nerve center, with the Centre of control in the thalamus. This is pure science even if it is told from a general point of view. Of this complex of reactions is that being the memory short and to long term as is part of the set of what is engraved in our inside. The logos or the global knowledge of the nature of things and comes as a magnet of these internal forces which flow into every human being in harmony with the universal forces and laws of creation.

3 meditation position since it was discovered for the first time this position of meditation, concentration and contemplation has been used by the different schools that guide groups in their spiritual advancement in meditation and concentration of forces and internal energy. In many countries of the world is something sacred, as in India, where SIDDHARTHA GAUTAMA better known with the Buddha 2,500 years ago, reached the master lighting and the domain of life. Many ups and downs became the development at different times and people who were his followers or students proclaim them gods. Like Jesus, the disciples spoke of his spiritual attributes and their great lighting, but none classifies

him as a God. It taught them was: "I am one of you, and what I do you can do so or bigger than me;" just follow the secret teachings that I give them and believe in them." The so-called Gospels Gnostics, declared apocryphal or hidden meaning, sublime, unknown to the general interpretation; Another way of rejecting what not was that he knew at the time or in the future. He sees our inside and concentrate these energies on every part of it makes us aware of its existence and at the same time, we activate our centers of communication with them. Once we become aware of each of them, from right to left, and got land through our body part by part, from right to left, and at the same time concentrating the energy accumulated through our breath, equip our body with a conductivity which is directed where concentrates it our conscious will. Passing our consciousness from the Earth until I have mysteries of our brain, from left side to right side of every part entails that the glandular system could be harmonized and function. Join our cellular system with its own independent evolution consciousness makes us take momentary control of some functions that we do not control our conscious part, which is that we can handle, as walking, look at all sides and features that we control at will. He repeated this exercise of visualization and concentration of energies below upward, of top-down, inside out, leads us, with proper preparation, to awaken our subtle spiritual side. Daily recurring prepares us until we could gather in the center of our system - thalamus - that subtle energy in the Center and they reveal the mysteries of all avatars want to know the humble heart. Return to be children once our subtle energies meet this Center, divine energy will harmonize with them and will manifest the true spiritual being of our God. That provide our bodies with feeding and respiration, which manufactured the material part of this and whose function is to keep stuff for the preparation of cells; the other part is the subtle part, with which our interior reproduces. The phenomenon of life is attracted to our material body as a fluid force that complements the processes of evolution. In the entire universe, there is this creative energy and it is possible that other galaxies State life equal to or with some features that follow the same laws of evolution of the human being.

Where there is a Lake, sea or body of water where the sodium and potassium are present, the being can exist also in some stage of development. Perhaps other reactions of the Alchemy of the same law have produced materials and similar energies that respond to the same impulses of our heritage of creation and are demonstrating its evolution beyond our reach and manifest intelligent life like ours, no matter if in a more advanced or delayed in development stage. The scientists wound brains trying to life occurs in outer space, where there is only water. Looking for intelligent life on other dimensions of cosmos, do the true expression of life on planet Earth or its origins, not salt water. Can find water in any planet and their conditions can be suitable, but while not is has of that the life smart that intend to develop in the space outside must be directed by a gland, like the thalamus in the human being. Its unique characteristics must be present in any system that tries to emulate the creation of human beings. The combination of sodium potassium, phosphorus and other minerals that react to the same conditions that occurred in the phenomenon of creation can be duplicated at a specific site with appropriate combinations of energies. If you meet these requirements and are looking for the appropriate place it is possible to repeat the phenomenon of creation. Our stagnation in knowledge is a problem of human behavior. During the last hundred years advanced rather than in all the earlier centuries. He only made of have challenged the authority of Governments and institutions still because of the sacrifice of lives of those illuminated that have left a legacy, that part is has revealed, us puts in a road where the history not must stop is. The heritage of human knowledge must not be stopped for any reason. Free Governments and scientific institutions with great respect for the heritage of the human being must organize their struggle to free the knowledge and that more beings engaged in the exploration of a decent future for humanity. Science and human progress in the future must be one of freedoms, considering the laws of divine creation that adorns the earthly existence. While these are coerced by false dictates of institutions that do not have a mastery of the scientific realities and that Governments have succumbed for their businesses and special interests. Stop the wars and the domain of the human being. The understanding and the globalization of knowledge must be one or the first goal of global security. The minds scientific is maintained to the expected of that arises some phenomenon that reveals this great discovery. The scientists who can dominate and understand this Act of creation will be at the gates of the great truths and revelations of the universe. One interpretation more real than life is itself would give a jump in space-time and would put us able to explain with simplicity the processes that Act on our lives. Speaking of lives, I mean everything that vibrates in the universe, from a black hole to the simplest cell. The tradition is usually penetrated the ancient texts, use imagination and inventiveness of languages and communication to accommodate ideas to new creations, adapting to new concepts accepted and invented by the idealists in different cultures.

It arrives believers to create in their minds the caches of the interpretations of each human being attached to a power to interpret the divine law.

176- The concept of eternal life

In the evolution of the human being from the first cell that manifested itself and evolved to an advanced state, it is revealed that a higher power was projected and exerted an influence on the interact in this first creation. This same force which clothes the entire universe and that vibrates in all their ends and contours, subtle Force that scientific, rare time are aware of its existence, related with the own existence and the universe. An intelligent force that simultaneously creates maintains a control that is what manifests itself, as if there is a control of all the attributes of the demonstrations, a subtle mantle which clothes all matter in the universe. What captures the human mind in its temporary cabin living, without power rise materially to other corners of the universe to contact with intelligences of other magnitudes are material or emotional as ours. We live in a world subject to laws that keeps us tied to the story that imposes knowledge accumulation and specialize in making keep a dependency of this. The reasons are diverse and the be fits them. Transcend concepts was a method old and sometimes present of mindsets that is rise to others levels of understanding and not is fastened to the interpretations accommodative. The concepts are created by great minds that carry an image of clarity to other beings so that they form a cloud of the imagination, per his personal mental capacity. With the imposition of controls the powers of interpretation, dictionaries and encyclopedias of learning, how to design knowledge is partially under the control of some institutions that shape the knowledge to unravel that is the method accepted and practiced, to the history that educates us. People that they are not in contact with these concepts, for example in the early stages of civilization, each culture created their ideas per the traditions that adapted the territories and cultures. To be straight with other more developed towns and could be less developed, confronting ideas, often created an amalgam of those accepted concepts and achieved an understanding accepted for future generations of these mixtures of cultures. The way these notions have matured, had overcome many changes, which many have lost their true and original meaning. It was a practice of the old schools to use petroglyphs to translate that knowledge. They had the certainty that the only thing available that does not diminish were materials at your fingertips. He practiced with all fibers, leather, wood, stone, primitive metal produced by primitive methods. What we shall now proceed to explain in words, we are looking for some method of cataloging the grouping of phenomena that occur outside our understanding and to which no human being nor science has reached to explain clearly. God catalogues the human being to all laws who were protesting outside the human understanding that is our legacy. In that address is oriented this knowledge and that can be of utility for all human being can address a knowledge more personal of the creation and them concepts that is relate, to what is da by sitting that is the Dios of each human being that accepts that denomination for their forces spiritual and of his soul and existence in this flat earthly. Form that was introduced this knowledge should be interpreted to the staunch and enthusiastic developers of those original concepts. The union of civilizations, the migration of ethnic groups, from the steppes of the North who brought their cultures and concepts matured for centuries. The migration of the inhabitants of those regions was one of survival by the calamities of the climate, the continuous wars between peoples, the scarcity of food and survival of races. The eldest of their resources was the livestock and agriculture, as well as fishing. Once this was reduced by the harshness of the weather, they become nomads in looking for new lands and territories where to settle. That is part of the civilization that grew out of the way in which people survived. This is a synthesis of how give an idea of the migrations of the early days. Maturation stages of beliefs and the formation of concepts of one God, was the best way to demonstrate what the eyes and experience was happening daily, for people who did not have a clear understanding of natural laws. He lived daily repetition of the births of new creatures, curiosity raise their minds to causes that were repeated daily to your around. New creatures have recreated them and under their care were developed with new and higher powers of expression. The acculturation of peoples, added more extensive ways of relating, they raised the imagination that has reached our era by way of contact with that knowledge. In that link in the story lost some of those concepts on how to meet the inside box of what meant form inner cover and publicize those legacies and on the maturation of the purity of those original ones. The elevation of Interior forces to communicate their emotional idea of what for them their way to visualize the divinity.

At some stage in the human relations concepts must have matured and become part of the story that is reported for periods of time. At which stages of evolution broke the magic of those original versions. Matured in some time of creating a concept of a supreme being or universal law where all content, repeated the laws of creation and the constant overcoming of the phenomena of the creation, was created in the human mind a mystery of how things work. The reason for the phenomena that repeated the creations of nature that is repetitive, that does not go away, whenever his life is renewed. The curiosity appears never-before-seen things before the astonished eyes of the only ones who realize these facts. You can interpret them in their natural environment and keep it in your memory as something that should be evaluated with new forms of cataloguing the events to their way of life. The maturity of those mental pictures is those that give us the certainty that somehow the concepts we have before our lives could mean different things. The heart of that question underlies them potential changes of what emerged as reality in the mind of which developed the concept and what is interpreter after them. Arises doubt lie or broken concept of reality. With that it replaces what is truth arises the concept of guilt of beings. What someone proposes as a reality of truth is the opposite thoughts or mental maps of other souls that coexist in a same room. The darkness of the concepts clothes the mind, which does not manage to overcome this stage of life should accept his guilt and beyond, and the concept of sin is born. Feel bad because another being projected an idea contrary to yours and that by acting outside its scope it has you be labeled a sinner, liar, etc. If you could recreate universal history in the right historical context, as they took place, every fact, every reason, detail to detail, take a sample from any period of history and can recreate out of the few realities, we surprised things that were left off the books. These observations are based on the curiosity of things which are discovered daily and slopes that overlook reality. He just feels an internal emotion upon hearing a word intonation, a whisper, a musical note, a vibration that surrounds us, can penetrate our being conscious and move to an unknown world of ideas and truths that sometimes words may not bring to reality. Glimpse into the passion of penetrating and transcend to the other side of the imagination, capture an imaginary cloud where you could recreate an energy strictly from a link in the creation, it would be bringing to reality a power discovers that cosmic area of the mind where talks be not being and seek knowledge from the common mind. Many developed beings have achieved this development and managed to transcend, or travel outside the usual material energy and leave their personal achievements and common aspirations, to capture this reality. The feeling that would give the trashing of their discoveries the notion of something in your inside them places in the border of the life with the death, both physical as mental. Accumulated for centuries terrors seize neurons, which processed the survival of matter, which succumbs to the pain and anxiety of being another victim.

180 - The Darkness

November 11, 2010

Darkness is defined as the absence of light.

For the active mind, which assesses the consequences of what they could touch materially and deny or not to attribute a function activated many of the phenomena for which there is no complete definition, who believes in our understanding a motion picture to which spend your efforts to find its real qualities of existence. I've been meditating on qualities or not qualities of darkness. Although it is not something tangible, it's a void of qualities that other active causes must travel to exist. There where the light does not penetrate conditions which are suitable for reactions that would be not activated if it were not for the absence of light. Although this is a failed assertion, it strikes me that although it is not a law stated and studied am a quality that motivates different reactions. In the absolute darkness, perhaps in the depths of the caves where light does not penetrate are created conditions of life and creation of species, plants that take place with the minimum amount of energy.

This shows that where the light does not penetrate fully reactions that occur only are possible when the darkness is present. Even the ability to view and the creation of bodies in species that develop are adapted to the darkness; or that, before the absence of light, the matter acquires others features unexpected that us put to think in others phenomena of her own creation. A reduction of light energy or rebounds of these produces phenomena that could not occur unless some degree of darkness makes it difficult for you complete and constant penetration on many reactions that occur in own matter. A seed that is exposed to the Sun constantly does not react the same way when it is under the Earth and only the darkness allows you to be reached by the amount of light that gives the right combination to make emerge in that plant life. Other conditions are combined in these reactions. The darkness is a condition that makes its counterpart, the light, is revealed. In our physical reactions, the darkness has a factor of protection which alerts us and makes us react to avoid trouble and avoid objects. The simple movement of a shadow makes us to react to the dangers, because we detect sensory event that follows and needed to respond appropriately alerts are activated in our protection system. Without these phenomena of shadows created by the dark these complex reactions had not developed in our physical system. It was discovered in our endocrine system and brain areas of neurons that respond to what is called fear or ability to react appropriately to dangerous situations. This is a condition that is part of our system of natural survival, which was developed in our system for survival since the beginning of the creation. In the case of situations of darkness, the way we react completely changes; areas of our system that are not common to light are activated. This proves that the darkness is an important factor in everything around us and all that has been created since the beginning of our existence. We could imagine under what conditions the human being and all creatures created had expressed and its evolution would have been quite another. The brain, which is a hot bed of chemical and physical reactions, where the current is handled, is such potential that just any disturbance to these functions creates what catalog of epilepsy. This is the case where large amounts of energy are released outside of its usual conductive channel and reactions are different. It is a download which goes to other areas where it was not intended and makes our central nervous system enters a crossing of signals and downloads that will be diverted to other nerve stem causing in many cases that the reactions are convulsions of the muscular system. By the only made of receive these downloads them muscles are contracted, in accordance with the capacity of energies that the invade it. Under these conditions is the sea of experiences of all types, such as hallucinations and States of euphoria where you live and feel different emotions that sometimes lead to the beings that suffer from them to enter a mental state of having received orders or mandates do things that before not perceived? Amounts of stories have been written of this phenomenon. During evolution, they have rescued stories of characters who have distinguished themselves in their performances to be epileptic and generally known cases point to religious reactions.

The shape of the brain rest and regain control of the internal faculties, blocking our conscious part and penetrate the subconscious to enter a more inclusive process of regeneration. Must do it in a climate calm, free of noise, and that all their processes are in harmony. It is precisely where the darkness must play an important role. For my remarks, it is a period where the light reactions are the minimum and internal processes are appropriate in the absence of the greater amount of natural light. If we process elements that invade us from cosmic emanations, it would be logical to think that the body and its components have been trained to receive and process certain rates of such emanations. If there are disturbances that decrease or increase such radiation, the body should regulate your receptive system to compensate for these variations to protect their States of internal processes.

Renowned scientists in the world give opinions that they are credible for most his colleagues. He argues the non-existence of an energetic substance; this substance does exist light and darkness; sodium and potassium-. What is manifest, the divine creation of the being, a deep reason that only try to explain God with materialistic theories and the creation of the universe. Whatever the evolution of the self and its attributes, given to the inherited human's intelligence, to recreate a divine intelligence. The evolution following laws cosmic of creation for them parts physical materials in the be, not you remove that spark divine that it associates to sources of emanations that not are perceptible or have could be explained by the science until this time. The entire range of human emotions that pervade reality to which every human being has access and enable it to experience it through their existence in this earthly plane and then completed his period of manifestation in a physical body to reincarnate in a new body to follow a spiritual evolution to reach a balance and greater spirituality. Human emotions may not be circumvented and achieved to the intonation of our divine inside that attract to us voluntarily and involuntarily, and that we project our being be towards other as a harmony of the own energy qualities that we use to attach our expressions, and are an achievement of our individual personality. These emotions that we do through our lives and that we can mature our intellect are not covered by any religious teaching, because it is only an attribute of the individual inner personality. They can inculcate us ideas of their potential and how to achieve it. Only the individual has the power to access it. It is a proven fact that the resurrection and reincarnation are still laws through what has made human history. It is also something real that the projection of the self through the time-space is possible. The return of the soul in a new body is the same process. Jesus taught this truth to the Jews and Apostles to enter the Kingdom there to be children again. Resurrection practices include great relationships in the history of religions. Peter raises Tabitha - Dorcas, Apostle female Priestess. Here repeats a story of an apostle resurrecting a follower to make known the mystery of the resurrection the same description of Jesus and Lazarus.

184- Reincarnation

The term refers to dating a process of return to repeat the incarnation or the new manifestation of attributes of existence of an entity that lived in a physical body at a previous relative. This word was known in the ancient teachings in the old and New Testaments. Attribute which was observed by ancient civilizations, and that the only ones who have documented it were Egyptians. It adopted as a knowledge superior, but not it managed to decipher all, which was declared anathema in the year 553 A.d. The way of resurrection was adopted to change the tone or meaning of the teaching of reincarnation applies to that Word. At the Council of Constantinople, in the year 553, by order of the Emperor Justinian, it was declared heresy and anathema. And the concept was removed from the catolic-cristians teachings. Not to be able to give a logical explanation and a Coptic and Greek term questioned the authorities face an existential reality that challenged their interpretation of the word and, at the same time, the process of resurrection that had taken to introduce the Christ of Saul in the Pauline writings. The statements already released records of the resurrection of Jesus as the area where the dead are returning was to them a threat that is wielded like a sword over their performances since Saul used it to justify their interpretations and personal preaching outside of Jerusalem about the resurrection of Jesus. The term resurrection covers only the phase where the soul detaches from the body freely or by control of the self that it dominates that faculty, which was common in the time of Jesus and was part of their teachings. In addition, was known centuries before and used in many writings and in the same Bible and Qumran writings.

Resurrection can be describing as the attribute of detached from the physical body, either voluntarily or involuntarily, and consciously return to the same physical - astral travel. This is the understanding accepted by the ancient religions that are practiced and taught at the school of the mysteries. Nobody had a teaching defined in terms of these natural phenomena, which are summarized in the attributes of being created; the soul or spiritual energy, abandons the body for short periods of time. For teachers of different civilizations intend to dispose of its knowledge, it is a practice of belief, to evoke the mysteries of his spiritual attributes. These as many ups and downs in different times, because their conditions of spiritual advancement by means of physical manifestations. This would be a direct proof of the causes which refers to which religions claim to be attributed to spiritual leaders claimed in each sect. The purpose is the same in each country or race that will proclaim a spiritual leader. It asceticism, the purification of the matter human, the domain of the energy of creation on the matter, gives a reality very different to what is known usually. The spiritual uplift is precisely what everyone yearns to understand inwardly. It is not simply a matter of desire it, to feel it. Many ancient civilizations in different cultures, adopted measures to induce their followers, using hallucinogen to give substance is feeling of elevation to a State of higher consciousness. History is a witness of those times, which are found those ways to conquer the minds of the unwary, are symbols of the history that dwells in the circles and the civilizations that succumbed to such practices. Behind the written history are the witnesses and the main actors of this plot.

The detachment of the spiritual energies is not a phenomenon unrelated to the reality of the physical laws of the human body, or the consciousness that we are gifted by the creation. Are conditions additional to what we are, natural to our being? If occasionally experiences this sensation while you sleep and body wake up relaxed and notions of new knowledge. The entrance of the energy of the soul in the body is a mystery to many researchers. You will be surprised that the subtle energy of the soul is simply a spiritual vibrational energy. A cluster of imperceptible vibrations traveling through the universe as a projection of the source that all scroll contains it. Attributes that it contains are attracted to this earthly plane by the laws of Astrophysics. On the planet, Earth where they inhabit the bodies of human beings, by adding everything that made up this planet in all its manifestations is the correlation of powers that are drawn or projected, towards areas of manifestation, in accordance with the stages of maturation of energy and the laws of attraction to it to create. The resurrection is a detachment of be physical of a quantity of energy that enables the functions of a body. The mass that contains a body, builds up a vibratory rate comparable to the energy of matter that consists of the. Based on molecular structures and Atomic components, that mass has a magnetic field. This in turn, is a mass of energy that attracts and drives out of its components, the waves of concordance with other materials that are expressed in the universe. Emissions from the laws of creation should be in unison a diapason of attraction and repulsion in the spectrum of the cosmic energy. The composition of the physical universe, as all that exists, is due to the influences of materials and the laws that dominate it. Science as such only displayed the effects captured by the human mind through the instruments used to guide the understanding of these laws to a level of human capabilities. We are the only ones that we see. Even if we penetrate the understanding of cosmic phenomena, the human mind will continue to grow by stages. The accumulation of data leads to a vision clearer of the phenomenon, and laws that interact with what catalogs life on this earthly plane. Possible awareness of this reality is an escape from the pure energy to a crude reality of nature. Travel from the confines of her purity to enter a phase of introspection of their reality of destruction or degradation that allows you to manifest and make them known outside your pure environment. An incongruity of the human mind to understand his own nature be a God power that never has been exceeded by their condition of being duped by realities that another project. The story will be the teacher; within its confines is the narration that never came to light. No one can hide a lamp on a high mountain, without affecting others, from the plains of the Temple of the Genezaret Lake, the historic mountain. Who found meaning to these words, not like death, because it returns to life and no one can erase his conscience, she will accompany it forever in his travels, back and back. Come a point of understanding of the sources of truth which embraces the universe. It will be a temporary illusion of the mental reality of the understanding. Those who accumulate data and experiences over eons of time to realize that the concepts are elusive, which is now a proven law, within a short time, will give a twist to their own dating that holds it in the annals accumulated as proof.

Reincarnate when ends your stage of life here on Earth and returns to the universal part of the divine essence, no matter the degree of advancement has been achieved in this earthly plane. A simple change of State of manifestation is catalogs of death. The mere fact of having to define these concepts, and when they stumbled across evidence that were already known and applied in different cultures, exposes some teachings that were based in an interpretation in history that does not tolerate scientific evidence.

The realm of the physical part of our inner Kingdom light is equipped with a Web of nerve networks that are at the same time internal censors where established a communications network where travel millions of impulses to the deepest parts of our physical structure. Its function is to receive information or signals that make possible the establishment of a constant communication and to take instructions to control different areas so that they supplement the materials necessary for the proper development of our body. In this way, if there is any nerve exposure, signals will be sent to areas where they are connected to alert you of the damage suffered or the materials needed to restore a specific function.

The process of reconstruction will be unleashed immediately. This Centre is in two levels of connections, internal and external, with the space outside our material physical spectrum. Creating gland receives all the impulses from the cosmic part and harmonizes them in the scheme of the physical manifestation of all creation emanating from the cosmic mind. It is only a link in the unknown. Physical pain is one of the features that cause any injury affecting a nerve terminal; It is one of the sensations more pronounced in our being. Intelligence that manifests itself in these reactions cannot be of material character exclusively, since signals will be sent to areas where they are connected to alert you of the damage suffered or the materials needed to restore a function. The dispatch of the impulses required for genetically cells and other aggregates are believed to repair the damage with the same characteristics of the previous. Why? Realizing these attributes account is where the emotions of the creation arise. Our system emotional is gripped in that harmony. But do not create it; we create harmony in the way we react to those impulses. We are equipped with a resistance to pain or any emotion that we seize. Many succumb to these experiences, but others are trained to resist the. Would mystery contain this attribute of our interior which makes us aware and sensitive of all our powers of expression? This attribute must be tied to an energy that provides us with life itself and must be part of the vital energy that flows from the cosmic God - our material body and spiritual. The reason why said this is because once the life ends this energy leaves the body and feelings disappear entirely, just like all the other attributes that distinguish us as created beings. The needs to attract energy sources giving us the qualities is part of that energy that vanish instantly interact. The secrets of the ancient Egyptians have been rescued by large illuminated quietly have been dedicated to collect those lessons and secretly for more than two thousand years, have kept the secret with the promise to make it known to those who are deserving of that enlightenment. Stakeholders should be sought in the initiatory and esoteric, schools as minds seeking lighting; did in antiquity simply type in their computers and the doors will open. Since initiated lamas in Tibet, in China, the gurus of the India and San José in California, Mexico and many other places around the world. In the era of the Egyptians, the people went in search of this knowledge and traveled great distances to meet the schools and be admitted. Where do they emanate all these instructions which are faithful copies of a previous demonstration? Why is duplicate in them, plants in their seeds, them humans in its evolution? - what role plays our imagination and our power of attract it cosmic to them flat physical and vice versa? When these faculties cease? - What is this mass of energy that we went and where is going to stop? - What law regulates the period between phenomenon's. The reorganization of the matter in so many varieties of expression that each day we surprise more, the diversity in that is manifest. The common mind is unaware of striking fluctuations in the phenomena that occur at every moment. Others are a large circle of expectation and everything that happens will be a new dawn in his life. The mysteries surrounding the evolution of the species for millions of years, keep us occupied, scientists, thinkers, mystics, and all that branch of knowledge that seeks to unravel the evolution of species, the largest chain of human knowledge, which is the carrier of its own information and that not has deciphered its content so far. Any form or explanation will end up forming another link in the chain in expansion simply supplementing the form in which it is projected. From the cosmic understanding to the lowest of what can be seen by the human mind expanding. Each being and mind is a world of individual creation. Everything is processed in accordance with the individual conception, is and will be part of a personal world; at the same time, will form part of the flow of the intelligence universal; you will not be rejected by the original source. Source of projection as well as the own universe expansion. In other worlds, whether this manifestation of intelligence, demonstrated to the expression of the human soul and its subtle emotions, which is what can be exported towards the components of the original creation. Intrinsically opens a world of possibilities for the interpretation of the original idea, although universal concepts of the same idea are shared. It will be influenced by the generality of the achieved knowledge. Knowledge that was saved in all possible ways: engraved in stone, parchments, secret keys, caves, writing, and so what we have today is the little that has survived to the harshness of nature and humanity. In them first data historical of the resurrection is mentioned to the Prophet Elias living in a cave, by any reason is shows the first example of resurrection in the history documented to the humanity. Elijah raises a child, son of a widow. Again, relates the resurrection where a cave is part of the panorama. When you declare a hidden truth in a symbol, does not care about the past: the symbol maintains the symbolism that was introduced as an educational, if the key would not be lost to its original interpretation. Once it manifests itself in an expression of the imagination powers invading the universe, the immediate truth is extinguished as the flame. With practice and spiritual evolution, we prepare so that each day go more deeply into making us aware of the divine nature. Everything being created goes through a process of visualization and awareness of how to transform transmute an idea into a real demonstration material. Everything that exists has materialized from this physical form. From imagination to creation, which are attributes of a human being? In this way, the being takes over the qualities and energies of the thing to occur and the plasma into a reality. That is part of the process of materializing the cosmic laws into something real. The human being has been consciously or unconsciously created for that. The imagination, the source of internal creation, is the reaction inside the range of energies that make up a current controlled and driven towards the harmonization of the own energies which manifested themselves in a creative impulse and the evocation of a desire that will take form and reality. We don't need to stop it at the time; this inner strength will continue acting and gathering the attributes of the thing until the human being want to physically perform the imagined plan. It is as emotional as in the physical form of expression. The seed will keep its essence creates until them conditions of manifestation are them correct so is give a good fruit, and if these is manifest in the object that is encoded in your code genetic. This process is identified with the reproduction in the wild, per the same laws. In the cell, the regeneration process produces cells that will be scheduled for a specific purpose and specific functions in areas which by its ID, by the internal system, shall be carried to the right place to have been intended. Controls in the human Centre is directed by the set of glands led by thalamus, pineal gland, neurons, brain membranes, central nervous system, the lobes of the brain matter and the range of materials and components that belong to this elephant of antiquity that inhabits the place of creation. This is the Grail of creation, the mystery of mysteries, the word incarnate, where they recreated the first cosmic vibration of creation, the divine logos, the lost Word. From the first time that this gland recreated the evolution in vibrations and took harmony in their environment, he created for himself a profile and a memory of all its components and was creating all the qualities that were necessary for their progress during its development. If your source depends on the evolution of the species, it must have arisen in unison in various forms and divisions which, in turn, kept the information like the seed, and the next manifestation would follow the same law of manifestation. The fact that science points in their recent findings about the qualities of this control center in our interior and the attributes that have been proven is equated with the ancient knowledge that dominated some of the enlightened ones of our planet. The first cell that arises and survived and was encapsulated in its environment due to them weather to which had that face is, you gave them attributes that the father-mother creator said in his involution towards it material, created from a space spiritual of projection towards the matter same. Meditation and concentration exercises highlight that this Center reacts to the concentration of energies that we project him and headed for the centers that compose it. The same when we concentrate on it and let it take control of our communication centers, noticed how to set an inner harmony and the energies of the spiritual-emotional travel by our body from heaven to Earth and Earth to heaven. Here's the interesting thing about this meditation. Much of the knowledge that has been accumulated by different civilizations and centuries attributed them to States of spiritual uplift which has not explained her conduct and continues to the present. Epileptic events are attributed to States of lack of control of areas of the brain system where are located the glands that make up the system and suffer downloads without control of energies that are supposed to flow normally for the proper functioning of the sequence in the human being and which dominate scientific knowledge is catalogued as rationally functional beings. Within the parameters set by the current science of human behavior and mental health are established parameters of conduct appropriate to what normally conforms to patterns of the manifestations of the own human beings. Starting of this premise, fits highlight that due to events that is have known from the antiquity and is have accumulated in the annals of them civilizations ancient, as the Egyptian and others that stand out in advances spiritual to level world, that have legacy a baggage cultural for the humanity, is should analyze this made with greater clarity and care. While it is true that these energies without control, cause features unusual phenomena, it is not less true that because of them manifests a reality that never has been considered since this picture. The ability of human beings to have a wealth of unknown functions and that science has tried to explain, and that today has been advancement towards this area of science. The ancient Egyptians dominated this knowledge and used it for their advances in all aspects of life. In the regions of Greece, the India, China and ancient Mesopotamia, transmitting a secret knowledge of the domain of this energy. The secrets of the library of Alexandria and many of the older files kept this secret knowledge, which the greater part was destroyed. Happens the same or similar with the proclamations of the Council of Nicaea, which most of this knowledge disappears from the known world stage after death and religious persecutions. The secret, underground should be the place to keep this knowledge away from the religious functions. That has been the experience of black history we have inherited. The control of this type of natural energies in the human being and that is a spiritual heritage of our divine creation. As I said, these energies control is an attribute our and with proper discipline and knowledge can be applied in the spiritual advancement of humans. The activity of the institutions that are dedicated to the creation of cause's fans currently remains active, trying to keep their domains and ensuring that the truth is not known. Current knowledge with truth, is forbidden, only that not can be sustained only by the Declaration of beings elected without a real reason. Such statements are valid by their councils and pawns in chess for your Board. Human intelligence is higher and it has crossed oceans and cultures to unravel the truth, we packed with models of a civilization, which crossed the Atlantic with those cramps and bumps than the imagination of the cloisters, organized for Brutes of this part of the world. Knowledge takes over human reason and cannot be stopped with threats or persecution. In the old schools that is called, of them mysteries taught and currently is teach them processes metaphysical that is have called for cataloging them of any form understandable as a branch of the know. Call school of the mysteries is not a designation of rarity or something out of the ordinary, is just one way of cataloging scientific knowledge is not at the level of ordinary people, and that was studied by those who had an advancement in education and preparation for the advancement of humanity. The search for the nature of God in our being and understanding of his true spiritual being in our soul should be the goal of all created beings. In greater or lesser degree, we will release ignorance which is us has inculcated by centuries.

Gaspar Edwin Pagan

196- Short biography

Gaspar (Edwin) Pagan Chevere

Born in Barranquitas, Puerto Rico January 07, 1945, son of Arcilio Pagan and Rosa Chevere

Some published titles The Last Witness- dedicated to my Brother Rafael Pagan, killed in an army Military base- being the last witness of Kennedys Dead. He was the last witness- one of the photographer of Kennedy body. 115 Persons that must see with Kennedys dead disappears in mysterious form.
Let's Play Dice with the universe.

When trying to enter the University in Puerto Rico in the year 68, it made me hard by the costs and the economy. Married in the same year, the obligations tied me to the daily routine. I could complete a course in mechanized accounting, with the base I got work at the Puerto Rico Medical center and other companies. Reaching positions of supervision, Assistant Manager at the international airport in Carolina, Puerto Rico. In 1985 melts my own company, Pelulleras DC Corp. Acting as CEO for 30 years until my retirement.
My constant internal search led me to join AMORC - Ancient and mystical order of the Rose and the Cross. Is a not sectarian organization, their bases are widely known by those search engines to level world as initiatory institution.
For mi surprise, must of greatest minds in history belong to AMORC. Einstein, Paracelsus, Leonardo Davinci and the strangest, Michael Nostradamus. For those who appreciate the inner mysteries I recommend AMORC.
I got great maturity in knowledge in this way as a student so far. The research on the history of Jesus, based on mystical experience is my passion. The search for details that have been changed by

institutions through history, bring them to light, understanding that the reliability is not in the historians.

Gaspar Pagan
January 28, 2017